The Milestones of Science

How We Came to Understand the Universe

JAMES D. STEIN

Prometheus Books

Essex, Connecticut

PB Prometheus Books

An imprint of Globe Pequot, the trade division of
The Rowman & Littlefield Publishing Group, Inc.
4501 Forbes Boulevard, Suite 200, Lanham, Maryland 20706
www.rowman.com

Distributed by NATIONAL BOOK NETWORK

British Library Cataloguing in Publication Information Available

Library of Congress Cataloging-in-Publication Data

Names: Stein, James D., 1941– author.
Title: The milestones of science : how we came to understand the universe /
   James D. Stein.
Description: Lanham, MD: Prometheus, [2023] | Includes bibliographical
   references and index. | Summary: "Comprised of riveting and readable
   stories from along the path of scientific discovery in the fields of
   Astronomy, The Earth, Matter, Forces and Energy, Chemistry, Life,
   Genetics & DNA, The Human Body, Disease, and Science in the 21st
   Century, author James D. Stein showcases the most noteworthy
   achievements of our species in a compelling and comprehensive way"—
   Provided by publisher.
Identifiers: LCCN 2022031125 (print) | LCCN 2022031126 (ebook) | ISBN
   9781633888487 (cloth) | ISBN 9781633888494 (epub)
Subjects: LCSH: Science—History—Popular works.
Classification: LCC Q126 .S75 2023 (print) | LCC Q126 (ebook) | DDC
   509—dc23/eng/20220705
LC record available at https://lccn.loc.gov/2022031125
LC ebook record available at https://lccn.loc.gov/2022031126

♾™ The paper used in this publication meets the minimum requirements of American
National Standard for Information Sciences—Permanence of Paper
for Printed Library Materials, ANSI/NISO Z39.48-1992.

Contents

Scientific Timeline

585 BCE   Thales predicts the first solar eclipse.

ca 350 BCE   Aristotle classifies five hundred known animal species.

ca 240 BCE   Eratosthenes measures the size of the Earth.

ca 225 BCE   Archimedes formulates the law of hydrostatics, to be known as Archimedes' principle.

164 CE   Galen becomes court physician to Emperor Marcus Aurelius.

1330   William of Occam publishes Occam's razor, the principle that one should prefer the simplest explanation.

1543   Nicolaus Copernicus publishes the heliocentric theory.

Vesalius publishes the first illustrated book on human anatomy.

1576   Tycho Brahe begins construction of the first astronomical observatory.

1590   Galileo's experiments with falling bodies refutes the Aristotelian theory of motion.

1609   Johannes Kepler publishes the first two of his Three Laws of Planetary Motion.

1628   William Harvey publishes his results on blood circulation.

1658   Jan Swammerdam discovers red blood cells.

1662   Robert Boyle shows that the pressure and volume of a gas held at constant temperature vary inversely.

1665   Isaac Newton begins the development of the theories of mechanics and gravitation.

Robert Hooke discovers cells in slices of dried cork.

1676	Anton von Leeuwenhoek discovers bacteria.
1735	Carl Linnaeus publishes *Systema Naturae*, outlining a biological classification scheme that is still used today.
1753	James Lind discovers that lemon juice cures scurvy.
1771	Luigi Galvani shows that frog muscle responds electrically when in contact with two different metals.
1779	Jan Ingenhousz discovers photosynthesis.
1781	Charles Messier publishes a list of "nebulosities" that are not to be confused with comets.
1785	Antoine Lavoisier disproves the phlogiston theory.
	James Hutton outlines several principles of modern geology.
	Charles Coulomb discovers that electric charge obeys an inverse-square law similar to gravitation.
1787	Jacques Charles shows that different gases expand by the same fraction for a given rise in temperature.
1794	Alessandro Volta shows that contact between two different metals will generate electricity.
1796	Edward Jenner develops a vaccine against smallpox.
1800	Humphry Davy discovers nitrous oxide, aka "laughing gas."
	William Herschel discovers infrared radiation.
1801	Johann Ritter discovers ultraviolet radiation.
1802	Thomas Young's double-slit experiment shows that light is a wave.
1803	John Dalton develops the modern atomic theory.
1807	Humphry Davy uses electrolysis to isolate potassium.
1811	Amadeo Avogadro hypothesizes that equal volumes of gas contain equal numbers of particles.
1820	André Ampère formulates the right-hand rule of electromagnetism.
1824	Sadi Carnot investigates thermodynamic efficiency.
1826	Johannes Müller discovers that any form of stimulation of the optic nerve is interpreted as light.
1827	Georg Ohm discovers the relation between electrical current flow, potential difference, and resistance.
1828	Friedrich Wöhler fabricates the organic compound urea.
1831	Michael Faraday discovers electromagnetic induction.

1832	Charles Babbage starts to design the "analytical engine," which would direct computation through punched cards.
1835	Gaspard de Coriolis discovers that the earth's rotation will deflect atmospheric and oceanic currents.
1837	Louis Agassiz finds evidence of glacial advance that he calls the "Ice Age."
1838	Matthias Schleiden announces the cell theory for plants.
1839	Theodor Schwann announces the cell theory for animals.
1842	Crawford Long uses ether to remove a neck tumor.
	Christian Doppler determines how a moving source changes the pitch of a sound; this is known as the Doppler effect.
1843	James Joule determines the mechanical equivalent of heat.
1845	John Couch Adams predicts the existence of Neptune; the next year Urbain Le Verrier would independently do the same.
1846	William Thomson (Lord Kelvin) obtains an estimate for the age of the Earth of 100 million years.
1847	Ignaz Semmelweis institutes antiseptic procedures to lower the incidence of childbed fever.
1854	John Snow uses statistics to analyze the spread of cholera during a London epidemic.
1856	William Perkin synthesizes mauve, the first artificial dye.
1858	Charles Darwin and Alfred Wallace read their papers on evolution before a London meeting of the Linnean Society.
	Rudolph Virchow states that "all cells arise from cells."
	Friedrich Kekulé presents his theory of chemical bonding.
1859	Robert Bunsen and Gustav Kirchoff make the first measurements of spectral lines of a chemical element.
1862	Louis Pasteur disproves the theory of spontaneous generation.
1864	James Maxwell publishes his first paper on the electromagnetic field.
1865	Julius von Sachs discovers that chlorophyll is responsible for the process of photosynthesis.
1866	Gregor Mendel publishes his work on inherited characteristics in pea plants.
1867	Joseph Lister uses carbolic acid to reduce deaths from postoperative infection.
1869	Dmitri Mendeleev constructs the periodic table.

1871	Ludwig Boltzmann starts work on statistical reformulation of the second law of thermodynamics.
1872	Louis Pasteur discovers anaerobic bacteria.
1876	Robert Koch demonstrates that bacilli cause anthrax.
1882	Walther Flemming discovers chromosomes and analyzes mitosis, the process of cell division.
1884	Svante Arrhenius postulates ionization to explain why solutions can conduct electricity.
1887	Albert Michelson and Edward Morley show that the speed of light is the same in two perpendicular directions.
1888	Heinrich Hertz detects radio waves.
	Edouard van Beneden discovers meiosis, the process of how sperm and ovum unite that is the basis of genetics.
	Friedrich Ostwald discovers that catalysts affect only the speed of the reaction, not the final result.
1895	Wilhelm Roentgen discovers X-rays.
1896	Antoine Becquerel discovers radioactivity.
	Svante Arrhenius raises the possibility that industrial production of carbon dioxide may produce a greenhouse effect.
	Eduard Buchner demonstrates fermentation outside the cell.
1897	J. J. Thomson discovers the electron.
	Christiaan Eijkman discovers that beriberi is caused by a dietary deficiency.
	Richard Oldham discovers that earthquakes generate two types of seismic waves.
1898	Martinus Beijerinck hypothesizes the existence of disease-causing organisms smaller than bacteria, which he calls "viruses."
1900	Max Planck's quantum theory resolves the "ultraviolet catastrophe."
	Karl Landsteiner confirms the correspondence between antigens and antibodies, and uses it to type blood.
1902	Ernest Rutherford and Frederick Soddy discover that radioactive decay results in the transmutation of elements.
	Ivan Pavlov demonstrates the existence of conditioned reflexes.
1905	Albert Einstein publishes the theory of special relativity.
	William Bayliss and Ernest Starling discover hormones.

1906   Ernest Rutherford's alpha-particle scattering experiments lead to the discovery of the nucleus.

1907   Bertram Boltwood uses radioactive dating to determine the existence of rocks more than 400 million years old.

        Emil Fischer synthesizes a chain of eighteen amino acids.

1909   Paul Ehrlich develops a cure for syphilis.

1911   Heike Kamerlingh Onnes observes superconductivity in mercury cooled near absolute zero.

        Thomas Hunt Morgan creates the first chromosomal maps.

1912   Alfred Wegener propounds the theory of continental drift.

        Henrietta Swan Leavitt works out the period-luminosity law for Cepheid variables.

        Max von Laue invents X-ray crystallography.

1913   Niels Bohr develops the "solar-system" model of the atom.

        Frederick Soddy hypothesizes the existence of different isotopes of the same element.

        Henry Moseley introduces the concept of atomic numbers and uses it to revise the periodic table.

1916   Gilbert Lewis introduces covalent bonding for chemical compounds.

1918   Harlow Shapley suggests that the Sun is fifty thousand light-years from the center of the Milky Way galaxy.

1921   Otto Loewi isolates the first neurotransmitter.

1924   Louis de Broglie shows that particles can behave as waves.

1927   Werner Heisenberg formulates the uncertainty principle.

1928   Alexander Fleming discovers penicillin.

1929   Paul Dirac predicts the existence of antimatter.

1931   Subrahmanyan Chandrasekhar predicts that a white dwarf star can exist only if its mass is less than 1.4 times that of the Sun.

1932   Carl Anderson discovers the anti-electron while studying cosmic rays.

1934   Leo Szilard conceives the idea of a chain reaction.

        Karl Popper defines falsifiability as the criterion for whether or not an explanation is a scientific one.

        Irene and Frederic Joliot-Curie produce radioactive isotopes not found in nature.

1935   Wendell Stanley crystallizes the tobacco mosaic virus.

William Rose isolates threonine, the last of the nutritionally important amino acids.

Albert Einstein, Boris Podolsky, and Nathan Rosen devise a thought experiment whose resolution will shed light on the nature of reality.

Hideki Yukawa develops a theory of the strong force.

1937   Hans Krebs discovers the citric acid cycle that is the primary source of energy for living organisms.

1938   Hans Bethe develops the theory of thermonuclear fusion to explain the mechanism that powers the Sun.

1939   Otto Hahn and Fritz Strassmann publish the results of the bombardment of uranium with neutrons; Lise Meitner recognizes that nuclear fission has taken place.

Linus Pauling publishes *The Nature of the Chemical Bond*.

1940   J. Robert Oppenheimer describes black holes.

1948   John Bardeen, Walter Brattain, and William Shockley develop the transistor.

1949   Melvin Calvin uses radioactive tracers in his studies on photosynthesis.

1950   Karl von Frisch deciphers the dances of bees.

1953   James Watson and Francis Crick decipher the structure of DNA.

Frederick Sanger works out the structure of insulin.

Stanley Miller simulates lightning in a prebiotic atmosphere and produces amino acids, the building blocks of proteins.

1955   Jonas Salk's polio vaccine is proven both safe and effective.

1957   Frank Burnet outlines the mechanism by which the immune system produces antibodies for specific antigens.

1960   Harry Hess develops the theory of seafloor spreading, which leads to the development of plate tectonics.

François Jacob and Jacques Monod discover the existence of messenger RNA.

1961   Edward Lorenz discovers the "butterfly effect" in weather forecasting, leading to the science of chaos.

1962   Marshall Nirenberg deciphers the first codon.

Rachel Carson publishes *Silent Spring*, documenting the effects of DDT on wildlife.

1963   Morris Goodman inaugurates molecular anthropology with biochemical studies of blood response to albumin.

Maarten Schmidt proposes that 3C 273 is an immensely distant radio source that will later be called a quasar.

1964   Murray Gell-Mann proposes that neutrons and protons are assemblages of three quarks.

1965   Arno Penzias and Robert Wilson discover the echo of the big bang.

1967   Allan Wilson and Vincent Sarich conclude from biochemical evidence that man and ape diverged 5 million years ago.

Jocelyn Bell and Anthony Hewish discover a rapidly rotating neutron star that will later be called a pulsar.

1968   Werner Arber discovers restriction enzymes.

1970   Howard Temin deciphers the mechanism by which retroviruses reproduce.

1971   Vera Rubin discovers that galaxies contain dark matter.

1973   Herbert Boyer and Stanley Cohen create the first genetically engineered bacterium.

1980   Walter and Luis Alvarez hypothesize that the collision of an asteroid with Earth led to the death of the dinosaurs.

Alan Guth proposes the inflationary theory of the Universe.

1983   Discovery of three "vector bosons" confirms the electroweak theory developed by Sheldon Glashow, Abdus Salam, and Steven Weinberg.

1984   Bradford Smith and Richard Terrile photograph a protoplanetary disk around the star Beta Pictoris.

1986   Karl Müller and Georg Bednorz create the first superconducting ceramic oxide.

1987   Allan Wilson and coauthors propose the existence of "mitochondrial Eve," a woman whose mitochondrial DNA is now part of everyone's genetic makeup.

1994   Alexander Wolszczan discovers a planet orbiting a pulsar.

Introduction

If one judges human accomplishment by the criterion of its contribution to the quality of life, science is almost certainly our most successful enterprise. Of course, we cannot overlook the enormous role played by technology—but without the guidance of science, technology would probably not have advanced much beyond what was achieved during the Age of Steam.

The remarkable effectiveness of science is due to a number of factors. It has taken several billion years of evolution to produce the brain and the five senses that *homo sapiens* possess, and science makes substantial use of all of these. Science consists of systematic knowledge of the physical universe using information obtained through observation or experimentation. The milestones of science fall into three categories—observations, experiments, and theories. Observation and experimentation use the senses, or artificial devices such as microscopes that extend the range of our senses. The systematization of this information is performed by the brain, or artificial devices such as computers that extend the range of our brain.

Theory is the systematization of information, and observation and experimentation are needed to both discover information and verify proposed systematizations. One of the reasons that science is so effective is that theory, observation, and experimentation are interdependent, and all play key roles. Science needs theory to systematize and enable us to predict, and also to decide which observations and experiments should be performed. Science needs observation and experimentation to gather data for systematization and to determine whether we have done a good job constructing theories.

Another factor that makes science so effective is that it is cumulative. No other branch of human endeavor builds so dramatically on what has already been accomplished. We often say that scientific knowledge increases

xiv THE MILESTONES OF SCIENCE

exponentially; although this is generally intended as a metaphor, it is quite probably the literal truth. The amount of knowledge being acquired is likely proportional to the amount of knowledge that already exists; if this is stated in the language of mathematics as a differential equation, its solution is that knowledge is indeed growing exponentially.

Science is also unique in that part of its toolbox is a methodology for determining errors. When I was young, I recall learning a verse explaining the meaning of the word "investigate":

> Investigate if you would know
> That what you think is really so.

Investigation is a critical part of science. If a scientific theory does not accord with the results of an investigation, no matter how beautiful the theory, it must be discarded.

CONNECTIONS

Many years ago, when I was first appointed to the faculty at California State University, Long Beach, I had an interview with the provost, a man of substantial erudition who felt that it was advisable to get to know the people whom the university was hiring. We talked for an hour, and one of the questions he asked was, "If you could take only one book to a desert island, which one would it be?"

I had just finished reading James Burke's *Connections*, and even though it had the advantage of temporal proximity to the asking of the question, I thought—and still think—that it would be my choice. For those who have not read the book, you have a treat in store for you. Burke managed to integrate science, history, biography, and technology into ten stories, each of which culminated in one of the seminal technological developments of the twentieth century, such as the computer.

I have used *Connections* as the model for the structure of this book. Instead of the stories culminating in a particular development, as did Burke's, each story is organized around a major scientific theme, such as the story of genetics and DNA. Burke, however, either knew or was able to research a good deal of the historical and political background that surrounded the events that he used to tell a story. My goal is somewhat different: to tell the story of a major scientific theme through its milestones—the theories, observations, and experiments that are the milestones themselves, and the people who were primarily responsible.

One thing that Burke made clear is that his choices were idiosyncratic—he mentioned that if he were again to write the story of the connections that led to the development of the computer, he might tell it in a completely different way. I don't feel that I possess Burke's exquisite sensibility for making those choices—but fortunately, I don't have to. Yes, I may err in omitting a great theory or experiment, or in including ones of lesser importance, but if I make sure to include the theories of Newton and Einstein in telling the tale of our investigation of the phenomenon of gravity, and include some of the important observations and experiments that went into the construction and verification of those theories, I won't have gone too far wrong.

Finally, Burke told his stories in chronological order. Burke wove history and technology together—sometimes history influenced technology, sometimes technology influenced history, but one could always see development and progress. One of the features that made *Connections* such an enjoyable read is that Burke had his choice of some of history's greatest personalities to insert into his stories. Most of the characters who appear in this book are—not surprisingly—scientists, but many of the people who appear in this book led lives every bit as interesting as the political figures and celebrities whose lives are often better known.

EXPERIMENTS AND OBSERVATIONS

According to the dictionary, an experiment is "a test under controlled conditions that is made to demonstrate a known truth, examine the validity of a hypothesis, or determine the efficacy of something previously untried." Experiments and theories form an important part of science, but so do the accidental discoveries that result from observations, such as Galileo's use of a telescope to look at Saturn or Leeuwenhoek's microscopic examination of a drop of water.

Observations guided by experiment also play a key role in everyday science. A while back I purchased—for the amazingly low price of five dollars—the 62nd edition (1981–1982) of the *CRC Handbook of Chemistry and Physics*. Weighing in at six pounds, this book is an exceptional value in terms of knowledge per pound, and stores the results of an almost inconceivable amount of observation. On page B-147, for instance, we find that the inorganic chemical compound sodium chloride, whose chemical formula is NaCl, has the synonyms "common salt" and "nathalite" (who knew?).

It has a molecular weight of 58.44, a density of 2.165, a melting point of 801°C, and a boiling point of 1413°C. These numbers had to be determined by experiment, for we certainly can't deduce these quantities from abstract theory. You have to weigh, melt, and boil the stuff—and that's what a lot of experimentation is about.

The *Handbook* is truly one of the great achievements of our species. Richard Feynman once said:

> If, in some cataclysm, all scientific knowledge were to be destroyed, and only one sentence passed on to the next generation of creatures, what statement would contain the most information in the fewest words? I believe it is the atomic hypothesis (or atomic fact, or whatever you wish to call it) that all things are made of atoms—little particles that move around in perpetual motion, attracting each other when they are a little distance apart, but repelling upon being squeezed into one another.

My feeling is that, if in some cataclysm all books were to be destroyed, and only one book preserved for the next generation of creatures, which book would best give those creatures the greatest opportunity to get up to speed in the shortest possible time? I believe it is the latest precataclysm edition of the *Handbook of Chemistry and Physics*. Although it might be a good idea to bundle it with Erik Oberg's *Machinery's Handbook*, just so we can rebuild our technology as quickly as possible.

There are observations and experiments—and then there are *great* observations and experiments. There are doubtless situations where it may be useful—or even critical—to know the melting point of common salt, but it's not one of the significant landmarks in the intellectual advancement of our species. The great discoveries of science *are* significant landmarks in our intellectual advancement. A great observation or experiment is one that plays a key role in a great scientific discovery.

As I see it, there are three ways in which an observation or experiment can play such a key role. The first is that an observation or experiment can be a great discovery in and of itself, or play a crucial role in hatching a great theory. This opens our eyes, enabling us to see things or conceive ideas that we may not even have imagined.

Second, an observation or experiment deserves to be called great if it is the vital piece of evidence that nails down an important theory. The important theories of science are like the foundations of a house—the house is built on the foundation, and a faulty foundation can easily lead to the collapse of the entire edifice.

Third, a critical step on the road to determining the way things really are is to determine the way things really aren't. The history of science includes numerous examples of scientists—and often great scientists—who look at the available evidence and piece them together into an erroneous theory. The discovery that a theory is wrong can be the critical step in determining how things really are. Sometimes a great observation or experiment decides between two competing theories, but it can also simply sound the death knell for an erroneous theory, clearing the way for a correct theory to rise from the ashes.

ELEGANCE

At our current stage of scientific development, most of the great theories are elegant. Although, as Richard Feynman noted, there may be a single great theory that defines reality—and such a theory would be the ultimate in elegance—Feynman also observed that reality could be like an onion, and we just keep peeling off layers. So far, though, elegance characterizes almost all the great theories in every branch of science. Relativity, the "big bang," and DNA replication are not only great theoretical descriptions of major phenomena, they are all unquestionably elegant. Sometimes the observations and experiments that lead to these elegant theories are themselves elegant— but sometimes not.

UNDERSTANDING THE COSMOS

I like how Carl Sagan described the Cosmos; he said that the Cosmos was all that was, is, or ever will be. It is a measure of the greatness of *homo sapiens* that in a few hundred years, located on a small planet circling an undistinguished star in the outlying region of the Milky Way galaxy, we have learned so much about our environment and ourselves. It is my hope that I tell this tale well enough that, in the case that we can only leave three books to future generations, this book may be thought worthy of accompanying *The Handbook of Chemistry and Physics* and *Machinery's Handbook.*

CHAPTER 1

Astronomy

Most people who become exceptionally interested in something can generally recall an experience that triggered their interest—but thanks to the internet, I can date mine exactly, even though it happened seventy years ago.

My family had moved from New York City to the suburbs. I was going to school, but my father worked in the city and he and I always got up early to read the *New York Times*. I would start with the sports section to catch up with the previous day's doings in baseball, and my father would start with the front section. We would finish and exchange sections—my father wasn't especially interested in sports, but the financial news was also in the sports section.

On September 1, 1951, my father called my attention to an event mentioned on the front page of the paper. There was going to be a partial eclipse of the Sun. My father knew it was dangerous to look at an eclipse directly, but the safety precautions of the day said it was OK to view them through exposed photographic film, and as my father was a photography buff, we had plenty.

Fascinated, I read the article. The eclipse was due to start at 5:25 p.m., would reach maximum coverage at 6:01 p.m., and end at 7:10 p.m. As I recall, there was also a black-and-white drawing on the front page of the *Times* indicating what was happening, and the paper may also have given the maximum percentage of the Sun's surface that would be covered. Fortunately, September 1 was a Saturday, and so my father was home to help me experience this.

I know the specific dates and times because I typed "solar eclipses visible from New York" into a search engine, and was directed to the National Aeronautics and Space Administration's (NASA) Goddard Space Flight Center

1

website, where I found a directory of *all* solar and lunar eclipses visible from New York from 1 CE to 3000 CE (which means Common Era, which has largely replaced AD in scientific texts). The fact that there is such a website is a tribute to both technology and science. I was born in an era in which my family had an encyclopedia as an immediate access to information, and if we needed more detailed information, we went to the library. I am still astounded by the volume and depth of information available through search engines from the comfort of one's home—and as will be discussed in a later chapter, the internet is reshaping not only what we know, but how we acquire that knowledge.

My father and I went out to the backyard a little after five o'clock and spent two glorious hours watching one of Nature's most spectacular displays. I remember that when we went back in for dinner, I asked my father how they knew when the eclipse would start and when it would end. My father told me that scientists study such things, and that's when I made a career change.

Until that day, my preferred career was a professional baseball player—despite the fact that I had evidenced nothing in the way of ability in this area. But the Yankees would have to find someone other than me to replace Joe DiMaggio in center field, as I was so enthralled by the ability to predict the exact timing of something so impressive as a solar eclipse that I decided to become a scientist.

The Solar System

THE SOLAR ECLIPSE OF MAY 28, 585 BCE

Whatever Thales of Miletus did of a scientific nature—and he did a lot—he was undoubtedly the first to do it, because Thales was the world's first scientist.

It is possible that many of Thales's achievements were the result of his travels, for he certainly visited Egypt and perhaps Babylonia. Consequently, he was familiar with many of the achievements of other cultures, and he undoubtedly was able to make use of the knowledge that they had accumulated.

Even if many of his achievements relied on knowledge gathered elsewhere, there is no question that some of his accomplishments were his alone. He was unquestionably the first mathematician. He was the first to conceive of mathematical idealizations, viewing lines as infinitely thin and perfectly straight, and he is also the first individual to state and prove mathematical theorems by formal arguments. Among his best-known proofs are that the diameter of a circle divides it into two equal parts, and that the base angles of an isosceles triangle are equal.

The prototype Greek intellectual, Thales was the first to blend astronomy and philosophy into the subject that is now called cosmology. He is the first person known to have asked the question, "Of what is the Universe made?," and to answer it without invoking elephants on the backs of turtles, or other mystical phenomena. Thales's answer, that the Universe was an infinite ocean in which the Earth floated as a flat disk, is obviously incorrect, but it is a fact that he asked the question and answered it in nonmythical terms, and that clearly marks him as a scientist.

Thales's greatest achievement, however, is the first accurate prediction of a solar eclipse. Nowadays when a solar eclipse is due, the news will be all over the internet, and the chances are extremely good that you can get a live video feed—especially if it's a total eclipse. Thales merely predicted that a solar eclipse would occur in the year 585 BCE. Ballpark estimates are a legitimate part of science, even if in this case the ballpark was pretty large. We can be certain of the date of May 28 because we are now able to determine that the only solar eclipse that occurred that year in that portion of the world happened on May 28.

It is known that the Babylonians were able to accurately predict lunar eclipses two centuries prior to Thales, but lunar eclipses occur much more frequently and the periodicity is easier to determine. Thales's prediction was historically as well scientifically significant, because the Medes and the Lydians were about to go to war. When the eclipse occurred, it was taken as a sign that the gods would not look favorably upon the war. The two sides signed a peace treaty and went home. The fact that the solar eclipse happened on May 28 makes the decision not to go to war the first historical event that can be accurately dated.

Thales was also the world's first all-around intellectual, combining his scientific inclination with an interest in philosophy and, perhaps surprisingly for a scientist, politics. Thales urged the Greek city-states to unite in order to defend themselves against Lydia. The Greeks, who were attracted to the number seven (the seven wonders of the ancient world), were later to construct lists of the seven wise men of ancient Greece. Thales was invariably placed first.

Thales was probably not the first person to be asked, "If you're so smart, why ain't you rich?" but he is the first one recorded to devise a brilliant retort. It is said that by his study of the weather he knew that the next olive crop would be a good one. He then purchased options on all the olive presses, and when the olive crop indeed proved bountiful, was able to obtain enough money by renting his presses to live comfortably for the remainder of his life. Having made his point (and his fortune), he turned again to science, philosophy, and politics.

THE HELIOCENTRIC THEORY OF THE SOLAR SYSTEM

Slightly more than five hundred years after the birth of Christ, Rome fell to the invading barbarians, beginning a thousand-year period known as the Dark Ages. One of the major factors in making the Dark Ages dark was an almost complete lack of scientific progress. A "media blitz" during the Dark Ages put forth the prevailing view that all the great questions had been answered: the philosophical ones by the Church, and the secular ones by the ancient authorities, such as Ptolemy and Aristotle. What questions arose that could not be answered were viewed not as puzzles to be solved, but as mysteries into which it would be blasphemous to delve.

One of the central problems was the geometry of the Universe. Since Christ had lived on Earth, it was obvious that the Earth was the center of the Universe. In the geocentric theory, the Sun, Moon, and planets revolved in circular orbits around the Earth, and the stars belonged to a great fixed sphere.

One obvious difficulty with this theory was that, from time to time, Mars, Jupiter, and Saturn would reverse their direction of motion in the night sky. This obvious discrepancy had to be patched up, and so the geocentric theory was modified by Ptolemy, who introduced the theory of epicycles, which assumed that these planets described circular loops within their basic circular orbits, somewhat akin to a person on the edge of a merry-go-round walking around in a small circle near the edge. The beautiful basic geocentric theory, which used simple circles to describe the motions of the planets, was now somewhat untidy.

It was easy to understand why, if the Almighty had placed the Earth at the center of the Universe, He had designed the Sun, Moon, and planets to rotate around it in perfect circles. Why the Almighty felt it necessary to use epicycles bothered some curious individuals, especially those who had been exposed to the philosophical principle known as Occam's razor, which asserts that the simplest explanation is usually the correct one. In 1514. Nicolaus Copernicus produced a small handwritten volume, entitled *Little Commentary*, in which the author was not named—but Copernicus sent it out to some of his friends. In it he stated seven axioms, three of which were pivotal in the heliocentric theory. These were that the center of the Universe was near the Sun, the annual cycle of the seasons was produced by the Earth revolving around the Sun, and that the epicycles were an artifact of viewing these motions from the Earth. In 1543, near the end of his life, he decided to publish these conjectures.

This theory was greeted with a literal firestorm of opposition from the Roman Catholic Church—its adherents sometimes being burned at the stake. Nonetheless, it made some important converts, one of whom was Tycho Brahe, a Danish nobleman who may be said to be the founder of observational astronomy. Brahe devoted much of his life to the problem of accurate astronomical measurements. He persuaded Frederick II, the king of Denmark, to underwrite the construction of an astronomical observatory, thus beginning a tradition of obtaining government grants for the study of pure science.

One other convert to the Copernican theory was Johannes Kepler, whose belief in the theory arose from a curious mixture of science, religion, astrology, and numerology. In order to accurately determine the orbits of the planets, Kepler spent more than twenty years refining Brahe's measurements. Kepler labored diligently to fit this data into a geometric scheme in which the circular orbits were determined by inscribing spheres inside the five Platonic solids (tetrahedron, cube, octahedron, dodecahedron, and icosahedron).

Try as he might, Kepler could not get the data to fit his hypothesis. He then made one of the great scientific decisions of all time. Rather than try to hammer data into a hypothesis, he abandoned his hypothesis to see if he could find one that fit the data. The result was Kepler's three laws of planetary motion, the first of which was that the orbit of a planet is an ellipse with the Sun at one of the foci. This result was one of the pivotal checks on the correctness of Newton's law of gravitation, from which all of Kepler's laws of planetary motion could be deduced.

Copernicus, Brahe, and Kepler came from entirely different backgrounds. Copernicus was actually a junior executive within the hierarchy of the Church, holding a position that today might be referred to as a deputy comptroller. His epic work might never have been published had he not been strongly encouraged to do so by Georg Rheticus, a young professor of mathematics and astronomy who lived with Copernicus for two years, and urged him to share his thoughts with the world.

Brahe was a playboy who literally partied himself to death. At a party in which he had imbibed a considerable quantity of wine, he followed the practice of the times and did not leave the table until his host did. As a result of the failure to relieve the pressure on his bladder, he became unable to urinate, and the resulting buildup of toxins within his system caused his death some eleven days later. Let it not be said that there is nothing practical to be learned from a history of science.

Kepler was a professor of mathematics who accepted court appointments in mathematics and astronomy, which would enable him to collect additional orbital data. One of Kepler's notable achievements was successfully defending his mother on a charge of witchcraft—the charge was dismissed on a technicality because the prosecution failed to follow the appropriate legal procedures with regard to torture. Sadly, echoes of this barbaric practice can still be found in a number of legal systems in the Western world. Kepler could also lay claim to the title of father of science fiction, as he wrote a book called *The Dream* in which he imagined voyages to other worlds than Earth.

THE LAW OF FALLING BODIES

When most people think of temptation, a course in geometry does not immediately come to mind. Galileo's father, however, was a mathematician, then as now a profession with substantial intellectual rewards, but— speaking as a math professor—not nearly as many earthly ones. As a result, he determined that his son should become a physician, then as now a profession desired by many parents for their children. Aware that the family proclivity for mathematics might be passed on to his son, he did his best to keep Galileo away from mathematics. His efforts were to be in vain, as Galileo heard a lecture in geometry, and then discovered a book on the mathematical and scientific achievements of Archimedes. Bowing to the inevitable, his father consented to let him study mathematics and science.

Galileo was the first scientist of the modern era to fully appreciate two important aspects of science: the necessity for accurate data in the construction of a theory, and the role of mathematics in providing a framework for physical laws. Galileo was never content to rely on the words of authorities when he could ascertain the truth for himself. Aristotle's claim that heavier bodies fall faster than lighter ones had been accepted for millennia. In a classic, though possibly apocryphal, experiment, Galileo dropped two cannonballs off the Leaning Tower of Pisa, one ten times heavier than the other. Both hit the ground at the same time. Galileo conjectured that a feather and an iron weight would fall at identical rates in a vacuum, an experiment that was performed by astronauts on the Moon in front of a worldwide television audience.

Galileo's work in mechanics prepared the way for Newton's revolutionary theories. Galileo measured the rate at which a ball rolled down an inclined plane. By varying the angle at which the plane was inclined, he was able to show that the distance traveled by a falling object was proportional to the square of the time that the object had been falling. This was a truly remarkable observation, as throughout his life Galileo was hampered by the unavailability of accurate means of measurement, and had to use his own pulse as a clock.

Other areas of scientific advance beckoned. When the telescope was invented, it was initially used for military and commercial observations. Galileo trained the instrument on the sky, and created telescopic astronomy. Among his many discoveries were mountains on the Moon, sunspots, the phases of Venus, and the four largest moons of Jupiter, which are now known as the Galilean satellites. These discoveries convinced Galileo that two pillars of Church dogma, the geocentric theory of the Universe and the perfection of the heavens, were erroneous. Despite the fact that Giordano Bruno had been burned at the stake in 1600 for similar heresies, Galileo was convinced that Church authorities would view the promulgation of his ideas with greater tolerance.

He would be sadly disillusioned. Even though he was an old man, nearly blind from imprudent observations of the sun, the Roman Catholic Inquisition brought him to trial. Showing far greater common sense than had Bruno, Galileo ostensibly recanted, although he is reputed to have said, in reference to the Earth, "And yet it moves!" as he was led off after the trial. Galileo was confined to house arrest for the remainder of his life, and died in 1642, the year in which Isaac Newton was born. The torch had been passed.

Although Brahe and Kepler were convinced of the truth of the Copernican theory, it was Galileo's telescopic observations that slammed the nails into the coffin of the idea that the Earth was the center of the Universe. The four largest moons of Jupiter could actually be seen, by anyone with a sufficiently powerful telescope, to revolve around Jupiter, clearly demonstrating that not all heavenly bodies revolved around the Earth. One would think that this overwhelming evidence would have convinced anyone. Then as now, belief continued to die hard, and while burning at the stake is no longer a recognized threat to a new scientific theory, many scientists have had to fight with as much determination as did Galileo to gain acceptance for their ideas. Nowadays, although the Catholic Church is not the threat to science that it was in Galileo's time, there are still religions that stand against scientific progress, and they have been joined by strong political factions in many countries—including the United States.

THE LAW OF UNIVERSAL GRAVITATION

The current COVID-19 pandemic pales into insignificance when compared with some of the plagues of the past. We see a lot of comparisons with the Spanish flu pandemic of 1919, which is estimated to have killed more than 100 million people—but in the four years from 1347 to 1351, the Black Death slaughtered approximately 40 percent of the population of Europe. Europe was periodically revisited by the Black Death, and it came again to London in 1665. Cambridge University, then as now a citadel of higher education, closed down for two years, and Isaac Newton, a young student, returned to his family home at Woolsthorpe. Possibly due to the fact that people didn't travel much in those days, the plague didn't strike country villages with the same ferocity that it had struck London.

There was, however, one extremely fortuitous consequence—Newton was left to his own devices, and had time to explore some of his ideas. As he put it, "in the two plague years of 1665 and 1666 . . . I was in the prime of my age for invention, and minded mathematics best of all." In those two years, Newton developed the law of universal gravitation—and set the model by which science in the future would be done.

Science consists of observations, experiments, and theories. Galileo had performed the experiments and made the observations, which resulted in a mathematical description of how bodies fall. Newton hypothesized that the force of gravity between two objects was directly proportional to the product of their masses and inversely proportional to the square of the distance between them. From this assumption, he was able to deduce Kepler's three laws of planetary motion. The aesthetic appeal of the simplicity of Newton's assumptions, as well as the demonstrable validity of the conclusions that could be drawn from them, caused as instantaneous an acceptance of this model for doing science as was possible at the time.

Much science—perhaps most science—follows Newton's guidelines. Observations and experiments are woven into a theory that makes predictions beyond the observations and experiments used to construct the theory. We then perform experiments and make observations to see if those predictions accord with reality.

Newton's law of gravitation changed the way we regard the Universe. Prior to Newton, it had been assumed that objects in the heavens moved according to different laws than did objects on Earth. The Universe was believed to function according to a design that man was not meant to understand. Newton replaced this mystical Universe with one that operated as a giant mechanism, whose inner workings could be revealed through rational investigation. This point of view is responsible for many of the most important advances of Western civilization.

Like many brilliant and creative individuals, Newton was beset with psychological problems. He was undoubtedly paranoid—fearful of criticism, he refused to publish many of his discoveries. His articles on calculus, a supreme mathematical tool, only came to light when the German philosopher Gottfried Leibniz, who independently invented calculus more than ten years after Newton had done so, announced his discoveries. The seeds of the *Principia*, which contained his ideas on universal gravitation, remained in his desk for twenty years. Only the impassioned pleas of his good friend, astronomer Edmund Halley (of Halley's Comet fame), persuaded Newton to publish it, and Halley himself had to finance the initial printing. History repeated itself; recall that Copernicus might not have published his heliocentric theory had not a good friend urged him to do so.

Back in 1999, *Time* magazine nominated Albert Einstein as its Man of the Century. I think they missed an opportunity to nominate a Man of the Millennium—and if they had, I would have started a write-in campaign for Newton.

I've always loved what the poet Wordsworth wrote, on seeing a bust of Newton.

> The marble index of a mind forever
> Voyaging through strange seas of thought, alone.

THE DISCOVERY OF NEPTUNE

We can be pretty sure that even before recorded history, the objects that appeared in the sky fascinated those who observed them. The Greeks were certainly among the first to make charts of the location of the stars, although there is evidence of other cultures also making such records. In doing so, the Greeks observed that certain luminous objects, which they called planets (from the Greek word for wanderer), moved rapidly among the stars. Five planets—Mercury, Venus, Mars, Jupiter, and Saturn—were visible to the naked eye.

The invention of the telescope accelerated the growth of observational astronomy. Two of the finest telescope makers, and consequently two of the finest astronomers, were the Englishman William Herschel and his sister Caroline. William was a superb lens grinder, and while he was at work, Caroline (who would become the first great woman astronomer) read aloud to him and fed him so that he could continue working. In 1781, he came across an object in the sky that he had never seen. Because it formed a visible disc in the telescope, rather than the mere point that a star would make, Herschel at first concluded that he had discovered a comet. Subsequent observations showed that the disc had a sharp edge like a planet, rather than the fuzzy edge characteristic of comets. When he had obtained enough data to calculate the orbit, he found that it was nearly circular, like a planet, rather than the elongated ellipse of a comet. The conclusion was inescapable: he had indeed discovered a planet, which was later named Uranus.

Its orbit was carefully determined according to the laws of Newton's law of gravitation. After several decades, though, discrepancies began to appear between where the planet was supposed to be, and where it actually was. These discrepancies were noted by the Astronomer Royal, Sir George Airy, who believed that they were due to imperfections in Newton's law. As a result, when he received a paper in 1845 by a Cambridge undergraduate named John Couch Adams concerning the orbit of Uranus, he paid it no attention.

History was to show that this was a mammoth error on the part of Airy. Adams, who was compelled to tutor at Cambridge in order to earn money for his tuition, had spent his vacation working on a radical theory: the orbit of Uranus was deviating from the calculated path because of the influence of an undiscovered planet. Adams had worked out the mass and location that this undiscovered planet must have had in order to cause the observed changes.

Adams was not alone. The Frenchman Urbain Le Verrier, working on the same hypothesis, also deduced the mass and location of the planet. Le Verrier, however, had luck on his side. Unlike Adams, he was an established astronomer. When Johann Galle of the Berlin Observatory sent him some preprints, Le Verrier wrote back to thank him and suggest that he look at a particular region of the sky. Le Verrier's luck continued, as Galle had just been sent new and improved maps of the area in which Le Verrier was interested. As a result, Galle became the first individual to see the planet Neptune.

The discovery of Neptune was a theoretical tour de force, and established beyond any possible doubt the validity of Newton's law of gravitation (although not much doubt existed at the time). Both Le Verrier and Adams went on to have distinguished careers as astronomers, but the discovery of Neptune was the high point for each.

The discovery of Neptune was not Le Verrier's first attempt at finding an unknown planet. Before tackling the problem of the discrepancies in the orbit of Uranus, he had noticed subtle anomalies in the orbit of Mercury, and attempted to account for them the same way, by postulating a planet he referred to as Vulcan (probably because it was even closer to the sun than Mercury, and would therefore have been hotter than Vulcan's mythical forge). No such planet was ever found, and the problems with anomalies in Mercury's orbit nagged astronomers until early in the twentieth century. The fault was not with Le Verrier's calculation, but with Newton's law of gravitation—as a young German-born physicist was to show in the twentieth century.

THE THEORY OF RELATIVITY

In 1905, Albert Einstein had probably the most astounding year any scientist has ever experienced. Upon finishing his doctorate, he found himself unable to find an appropriate job in the academic world, so he took a job with the Swiss Patent Office in Berne. During the day, he was a civil servant, scrutinizing patent applications for such ordinary devices as an improved gun, and a new use for alternating current. During the evening, he would sometimes stop at the nearby Cafe Bollwerk for a cup of coffee and conversation. Somehow, amid the responsibilities of a nine-to-five job, he managed to write three papers. Two of these papers led directly to Nobel Prizes; the third paper was merely brilliant.

Einstein is best known for the theory of relativity, a brilliant restructuring of how gravity operates. When Einstein first started thinking about the ideas that were to lead to the theory of relativity, he performed a famous thought experiment. Suppose that, at the exact moment that a clock struck noon, you were able to ride away from the clock on a beam of light. Because light itself carries the information from the clock about what time the clock shows, you would forever see the clock as showing noon. Time stands still if you happen to be traveling at the speed of light. The corollary is that time itself is not absolute, but depends on the observer.

The relativity of time appeared in Einstein's 1905 paper, "On the Electrodynamics of Moving Bodies," which lay at the core of the theory of special relativity. But Einstein was to go further. A decade of work culminated in the theory of general relativity, Einstein's reformulation of the Newtonian Universe.

In Newton's Universe, space and time are independent and absolute quantities, which can be measured by any observer. At any given moment, the Universe is a snapshot of a spatial stage, and the masses are the actors moving about this stage in relation to one another. The Universe according to Newton is an unfolding motion picture.

In Einstein's Universe, space and time are interlinked to form a four-dimensional geometrical structure called spacetime. The shape of spacetime determines how objects move; conversely, the objects themselves determine the shape of spacetime. The Universe according to Einstein is a geometrical entity in which space, time, and objects are all indissolubly related to one another. The differences between the Universes of Newton and Einstein are not apparent in everyday life, manifesting themselves mostly when objects travel at very high speed or are exceptionally massive. The first demonstration of the correctness of Einstein's reformulation was made in 1919, when Sir Arthur Eddington led an expedition to Africa to view a total eclipse of the Sun. Only then would it be possible to make the critical measurements of the motion of the planet Mercury to decide whether Einstein had supplied the answer to a discrepancy that had appeared in Newton's theory. In extreme situations such as described previously, Einstein's theory was proven correct, and since then Einstein's theory has passed every test with flying colors.

Einstein was unquestionably the most famous scientist who ever lived, and arguably the most brilliant. Unlike the moody and paranoid Newton and the gloomy and pessimistic Darwin, he was a warm and charming humanitarian. Unlike many celebrities, he was aware of both his strengths and limitations. As one of the leading spokesmen of the Zionist cause, and certainly the most famous, he was the first to be offered the presidency of Israel when that nation came into being. He turned it down, saying that he had no great understanding of human problems.

Einstein was well known for his proclivity to phrase statements about the Universe by ascribing various points of view to God. When he was asked in 1919 how he would have felt if the measurements made during the total eclipse had not confirmed his predictions, he replied that he would have criticized God for a bad job in designing the Universe. Einstein's view of the probabilistically based subject of quantum mechanics is certainly best summarized in his well-known quote that "God does not play dice with the Universe." Finally, his good friend Niels Bohr, with whom he used to take long walks and discuss physics, grew rather tired of these pronouncements, and told him, "Stop telling God what to do." It took someone of the genius of Bohr to achieve a put-down of Einstein. Nonetheless, the feeling that the scientific community has toward Einstein may have been best expressed by Jacob Bronowski in *The Ascent of Man*: "Einstein was a man who could ask immensely simple questions. And what his life showed, and his work, is that when the answers are simple too, then you hear God thinking."

Stars

THE PERIOD-LUMINOSITY CURVE
OF THE CEPHEID VARIABLES

Among the questions that have undoubtedly been asked at all times and in all cultures is, "How big is the Universe?"

By the beginning of the nineteenth century, the construction of telescopes had improved to the point where it was actually possible to detect parallax, which is a type of relative motion of nearby objects against a fixed background. You can experience parallax for yourself if you hold a finger in front of your nose, and then look at the finger with one eye closed, then the other—the background shifts relative to your finger. Using this technique, Friedrich Bessel concluded in 1838 that the distance of the star 61 Cygni was more than six light-years from Earth. This discovery enlarged the Universe substantially, as even so great an authority as Newton had estimated that the stars were no more than two light-years from Earth.

Throughout the remainder of the nineteenth century, the measurement of parallax was the leading-edge technique for determining distances to the stars. Telescopes became even more powerful, and smaller parallaxes could be detected. This pushed the threshold of the furthest measurable stars to several hundred light-years. Because the distance to most stars could not be measured, astronomers naturally suspected that the Universe was much larger, but how much larger was still anybody's guess.

In the first two decades of the twentieth century, several different events combined to make possible more accurate measurements of the size of the Universe. The first was the definition by Ejnar Hertzsprung, a Danish astronomer, of the concept of absolute magnitude of a star. Previously, the magnitude of a star (now called the apparent magnitude) was a measure of how bright the star appeared. This is a function of the star's intrinsic brightness and the distance of the star from Earth. Hertzsprung suggested that one could compute how bright a star would appear if it were at a standard distance. As a result, there was a simple equation connecting three numbers: absolute magnitude (Hertzsprung's number), apparent magnitude, and distance from Earth.

The distance of a star could be computed if both the apparent magnitude and the absolute magnitude of a star were known. Astronomers had been measuring apparent magnitude for years, but the difficulty lay in computing the absolute magnitude of a star. When astronomers looked at a dim star, there was no way to tell if they were looking at a dim nearby star, or an extremely bright one dimmed by distance.

Henrietta Swan Leavitt was an astronomer working at Harvard University. She was particularly interested in a class of stars called Cepheid variables. These stars, originally discovered in the constellation of Cepheus, brightened and dimmed in an extremely regular fashion. Leavitt's great discovery was that there was a mathematical relationship between the absolute magnitude of a Cepheid variable and the length of the period of that variable—how long it took to go from brightest to dimmest and back to brightest again. These periods were obviously easy to measure; one simply timed them. The distances of several Cepheid variables had actually been computed by then, and so it was possible to use these measurements to calibrate the Cepheid yardstick. Several years later Cepheid variables were discovered in collections of stars that would later be known as galaxies, and for the first time it was realized that the Universe was at least millions of light-years in diameter.

The Cepheid variable yardstick is still the only one whose accuracy is acknowledged by the astronomical community as a whole. One cannot use this technique for distant galaxies, though, as it is impossible to make out individual Cepheid variables in such galaxies. One of the latest proposals is to update Leavitt's period-luminosity law for Cepheid variables by trying to find a different observable type of star for which there is a correlation between brightness and distance. The current candidate is the Type Ia supernova, which is visible over huge distances. Astronomers are currently trying to work out a law for Type Ia supernovas analogous to the period-luminosity law for Cepheid variables. If such a law exists, it would bring a realistic determination of the size of the Universe within reach.

Leavitt's life serves as a good indicator of the second-class status (if that) held by women during her lifetime. She eventually graduated from what is currently known as Radcliffe College with a certificate that stated had she been a man, she would have received a Bachelor of Arts degree. Harlow Shapley, who was one of the leading astronomers of the times, wrote to her superior at Harvard Observatory that, "Her discovery of the relation of period to brightness is destined to be one of the most significant results of stellar astronomy, I believe."

Leavitt died in 1921, but despite her contributions, the news of her passing was not known to a number of prestigious scientists, including Gösta Mittag-Leffler, who wrote a letter to her in 1925 in which he stated,

> Honoured Miss Leavitt, What my friend and colleague Professor
> von Zeipel of Uppsala has told me about your admirable discov-

ery of the empirical law touching the connection between magnitude and period-length for the S Cepheid-variables of the Little Magellan's cloud, has impressed me so deeply that I feel seriously inclined to nominate you to the Nobel prize in physics for 1926.

It's hard to believe that the death of someone whose work was held in such esteem would have gone unnoticed—and, in keeping with her graduation certificate, that death probably would not have gone unnoticed had Leavitt been a man.

FUSION AND THE LIVES OF STARS

How does the Sun supply the light and heat necessary for life on Earth? With the discovery of the laws of thermodynamics, it was quickly seen that chemical burning, such as takes place in coal, was far too inefficient. In 1854, the German physicist Hermann von Helmholtz considered a subtler mechanism for producing heat: gravitation. The kinetic energy of particles falling toward the center of the Sun could be converted to radiation in accordance with the laws of thermodynamics. This would power the Sun for 25 million years.

Unfortunately, geologists had supplied convincing evidence that the Earth was at least hundreds of millions of years old, and so gravitational conversion of mechanical energy to heat radiation was clearly not the answer. The problem would remain unsolved until the start of the twentieth century, when additional sources of energy were discovered within the heart of the atom itself. Einstein's famous formula $E = mc^2$ demonstrated that the Sun clearly had more than enough mass to generate energy for billions of years, providing that a mechanism to convert energy with sufficient efficiency could be found.

The individual who figured out the conversion technique was Hans Bethe, a German physicist who escaped from Nazi Germany and ended up at Cornell University. Bethe was familiar with nuclear processes, and he had also read Arthur Eddington's conclusions that the temperatures in the interiors of stars had to be on the order of hundreds of millions of degrees. Using these results, Bethe was able to postulate a process whose result was the squeezing together of hydrogen nuclei to form helium. The helium resulting from this "fusion" process weighed less than the hydrogen that formed it, and Bethe was able to show that the missing mass was converted to energy in accordance with Einstein's formula. The Sun was a giant furnace, converting

4,200,000 tons of mass to energy every second. Because of the huge size of the Sun, the Sun could continue to radiate heat and light for billions of years.

This process, which goes on in all stars, results in a delicate balance between the star's radiation pressure, which makes the star expand, and its internal gravitation, which makes the star contract. At approximately the same time, the Indian astronomer Subrahmanyan Chandrasekhar determined that the outcome of this battle depended on the initial size of the star. The very small stars gradually exhaust their fuel and cool to a dull red. In larger stars, whose initial mass is less than about 1.4 times the Sun's mass, the rate of burning is faster and gravitational contraction is stronger. Such stars end their lives as white dwarfs—extremely hot, but very tiny.

In the ensuing years, a more detailed study of the fusion mechanism in even larger stars has been developed. After a star has burned its hydrogen to helium, it contracts, and this contraction heats up the central core further. This added heat enables helium to be fused to carbon. When the helium has been consumed, the star contracts further, becoming hot enough to fuse carbon to oxygen. And so it continues, with added contraction enabling oxygen to be fused to neon, then silicon, sulfur and, finally, iron. These events occur at an ever-faster rate. When the core of the star becomes iron, it can no longer fuse. The long battle between radiation pressure and gravitational contraction is won by the latter, and the star collapses and rebounds in one of the most dramatic events in the Universe—a supernova explosion.

The lives of the stars are long, and man has only been observing the Universe intensively for four hundred years. However, there are so many stars in the Universe that, given the observing power of today's telescopes, sooner or later a really interesting event will be observed. "Sooner" came in 1987 with the discovery of Supernova 1987A, the first nearby supernova to be observed in more than three hundred years. All the important predictions of the theory were upheld. The theory of the lives of stars is of profound importance to us, for it postulates that all the heavier elements, from the calcium in our bones to the iron in our blood, are formed in supernovas. We are, in a very real sense, intimately connected to the Universe: our very lives are possible because of the violent deaths of stars.

BLACK HOLES, QUASARS, AND PULSARS

As anyone in the advertising business can attest, good packaging can make any product more attractive, even a product like an arcane physics concept. The term "black hole" was coined by John Wheeler to describe a situation that at first blush seems unbelievable. After an extremely heavy star becomes a supernova, it leaves behind a remnant that has so much mass packed in so small an area that there is no effective barrier to the force of gravitational collapse. This procedure, initially conceived by the English astronomer John Michell in 1783, was first described in full detail by J. Robert Oppenheimer before he was called away from theoretical physics to head the Manhattan Project, the top-secret World War II project to develop the first atomic bomb. Like the Energizer Bunny, the supernova remnant just keeps going and going and going—until the gravitational force is so strong that not even light can escape. Then it is gone—from the Universe.

Initially the idea of a black hole attracted a good deal of interest in the astrophysics community, but when no one could supply a candidate, that interest dwindled. Then, in 1963, the astronomer Maarten Schmidt made a discovery that was to incite new interest in the possibility of black holes.

Schmidt was studying an astronomical object known as 3C 273 with the huge Mount Palomar telescope. 3C 273 looked like a star, but had a spectrum unlike that of any known star. In a burst of insight, Schmidt realized that it would be possible to attach meaning to the spectral lines of 3C 273 if one assumed that 3C 273 possessed an extremely high red shift. The only way objects possess a high red shift is if they are receding at an enormous velocity. In view of the Hubble relationship between recession velocity and distance, this meant that 3C 273 was an astounding 2 billion light-years away. The only mechanism that astronomers could imagine that would make an object that far away appear that bright was a black hole with the mass of billions of suns, continually gobbling matter and converting it into radiation. Other objects similar to 3C 273 were discovered and given the name "quasi-stellar radio sources," from which the term "quasar" is derived.

Four years later, British astronomer Anthony Hewish had assigned his graduate student Jocelyn Bell to study quasars. She detected an extremely unusual radio signal from one of them, a pulse with metronomic regularity. Initially, Hewish could not conceive of a natural mechanism to account for the regularity of the signals, and was considering the possibility that he and Bell had stumbled upon a signal beacon from an extraterrestrial life-form. Not entirely in jest, they referred to the object as LGM-1, where LGM stood for "little green men."

After several more months, four more such objects had been discovered. Hewish was able to abandon the "little green men" theory when, on further reflection, he realized that the signal could be a radio pulse from a rapidly spinning neutron star. Neutron stars are the residue of the explosion of a supernova, but the supernova is not quite massive enough to degenerate into a black hole. The star is so massive that the electrons and protons of its atoms are forced together and annihilate each other's electrical charge, leaving only electrically neutral neutrons. The spinning star acts like a cosmic lighthouse, with the radio beam regularly flashing past Earth, and this was the signal that Bell had detected. The term "pulsar" is used to describe the spinning neutron star which emits radio pulses.

It should be observed that while there is much inferential evidence for the existence of black holes, the riddle of the quasars has not yet been unraveled to the satisfaction of all astronomers. In 1974, Joseph Turner made careful studies of the rotation rate of a neutron star. Einstein's theory of relativity predicted that such an object should radiate gravitational waves, and the energy loss should slow down the spinning of the star. Turner's studies confirmed this, killing two birds at once: he had demonstrated the existence of both neutron stars and gravitational waves.

Oppenheimer's brilliance as a physicist was overshadowed—if that is the correct word—by his experience in heading the Manhattan Project, which some believe would not have succeeded with any other physicist in the position Oppenheimer occupied. Oppenheimer was not only brilliant, but by all accounts immensely charismatic—and his involvement in the Manhattan Project persuaded others to sign on. After the conclusion of the war, his influence faded as a result of concerns about his association with members of the Communist Party.

The Universe

THE STRUCTURE AND DIMENSIONS
OF THE MILKY WAY GALAXY

Not until the twentieth century did we realize how large the Universe really is. We have also made the fascinating discovery that there is a simple correlation between the size of the Universe and the age of the Universe.

Although the telescope was first invented and used for astronomical purposes in the seventeenth century, observational astronomy only became a popular pursuit in the eighteenth century with the invention of improved lenses. One of those who became interested in this field was Charles Messier, a Frenchman who was the first person to spot Halley's comet when it returned, as Halley had predicted, in 1758. This inspired him to spend his life searching for comets.

Unfortunately, there were numerous objects in the sky that appeared in a telescope to be as fuzzy and blurry as comets, so Messier decided to keep track of these objects so he would not mistake them for comets. He found and catalogued more than a hundred of these objects. Some would later be revealed as mere wisps of dust, but others would be vast collections of stars that Messier's telescope was simply too weak to resolve into individual stars. Messier found some twenty-one comets during his lifetime, none of which were memorable. However, many of the "nebulosities" he cataloged have been of immense astronomical importance.

The thirteenth entry in Messier's catalog is known to astronomers as Messier 13, or M13 for short. It is actually a huge cluster of more than a million stars, now called the Great Hercules Cluster because it appears in the constellation Hercules. Between 1915 and 1920, Harlow Shapley of the Mount Wilson Observatory made a study of globular clusters similar to M13. The 100-inch telescope at Mount Wilson was so good that many of the individual stars in the clusters could be resolved.

Fortunately, several years earlier Henrietta Swan Leavitt had worked out the period-luminosity curve of the type of stars known as Cepheid variables. The period-luminosity curve enabled one to compute the distances of these stars. Shapley decided to work out the distances of the globular clusters.

The globular clusters were not evenly distributed over the sky, as might have been expected if the Sun were at the center of the Milky Way galaxy. Instead, the globular clusters were distributed roughly spherically, and the center of the sphere was located somewhere in the constellation Sagittarius. Since the Milky Way galaxy seemed to be symmetric about a central point, and since there was no logical reason for Shapley to believe that the globular clusters were not also distributed symmetrically, he suggested that the center of the sphere of globular clusters was also the center of the Milky Way galaxy.

Having computed the distance of the globular clusters, Shapley was thus able to compute the distance to the center of the galaxy. In 1918 he proposed a model of the galaxy in which the distance from the Sun to the galaxy's center was approximately fifty thousand light-years. This distance was far greater than anything previously suggested, and was actually a little too large; the currently accepted figure is thirty thousand light-years. Previous underestimates occurred when astronomers had assumed that, because the stars in the Milky Way seemed equally bright in all directions, the Sun was at the center of the galaxy. Shapley pointed out that dark dust clouds obscured the bright center of the galaxy, leaving us only able to see the stars in our immediate neighborhood. This was later confirmed by radio astronomy.

Shapley's work also continued the process, which had begun with Copernicus, of demoting the Earth from the central position in the Universe. Not only was the Earth not the center of the solar system, the Sun was not even in the center of the Milky Way galaxy. The Milky Way galaxy has been shown to have a spiral structure and the Sun is in one of the arms, about two-thirds of the way from the center to the edge.

And that's a good thing. According to current theory, the galactic habitable zone—that portion of a galaxy where intelligent life is most likely to form—is some distance from the center of the galaxy, which contains the greatest density of supernovae and other energetic cosmic events, the radiation from which is capable of sterilizing planets some distance away. It may be more exciting to be where the action is in the center of a galaxy, but supernovae are—as they said of war in the 1960s—harmful to children and other living things.

THE BIG BANG THEORY

In the 1920s, Edwin Hubble began to make measurements of the velocities at which the galaxies were moving, and cosmology—the study of the Universe as a whole—began.

When Hubble measured the velocities of the galaxies, he discovered two extraordinary facts. The first was that almost every galaxy was moving away from us. The second was that the speed at which galaxies were moving depended in a straightforward fashion on how far away they were from us—if galaxy A was twice as far away as galaxy B, then galaxy A was moving away twice as fast as was galaxy B. This implied that the Universe was expanding.

By the early 1950s, two theories had been proposed to account for this expansion. In the steady state theory, the Universe looked the same at all times, past, present, or future; the Universe had no beginning and no end. Although the galaxies were rushing away from one another, new matter was being created at a rate that ensured that the galactic density would remain the same at all times. In the big bang theory, however, the Universe had a well-defined beginning in a gigantic explosion, and the galactic density—the number of galaxies per cubic light-year of space—would decrease as time advanced.

To decide between the two theories by waiting a couple of billion years and measuring the galactic density was obviously unappealing. Fortunately, there was another way to decide which theory was correct. The big bang theory predicted that the echo of the giant explosion from which the Universe began could be detected by sufficiently sensitive radio telescopes, whereas the steady state theory made no such prediction.

In the early 1960s two radio astronomers, Arno Penzias and Robert Wilson, had been recruited by Bell Laboratories to modify a radio antenna to bounce signals off the Echo satellite so that the antenna could send and receive microwave transmissions from the recently launched Telstar satellite. Once the modifications had been made, Penzias and Wilson would be allowed to use the antenna for radio astronomy.

Try as they might, Penzias and Wilson could not eliminate an unexplained background noise that seemed to come from everywhere. It was a challenge to their skill as engineers, and they spent over a year doing everything possible—cleaning out roosting pigeons (and the residue of the pigeons' tenancy), even rebuilding the antenna. They were about to dismiss the signal as spurious when Penzias was informed of a paper by a Princeton astrophysicist predicting that the echo of the hypothetical big bang might be detected with a sufficiently sensitive radio antenna. Penzias and Wilson called up Princeton, and the two groups ended up writing back-to-back papers in the *Astrophysical Journal.* The paper by Penzias and Wilson described the technical nature of reconditioning the antenna, the problems, and the leftover background noise. The paper by the Princeton group pointed out the possible interpretation that the signal was the leftover echo of the big bang.

Rarely has a plausible theory succumbed to experiment so quickly. Within a short time, the steady state theory was dead, and the birth announcements for the Universe were sent out.

In 1978, Penzias and Wilson were awarded the Nobel Prize for physics. Although great discoveries have often come about unexpectedly, to be awarded a Nobel Prize for accomplishing something when you (a) were trying unsuccessfully to do something else, and (b) needed outside help to make you aware of what you had accomplished, must set some sort of record for intellectual good fortune. If you want to win a Nobel Prize, your best chance is by being willing to get your hands dirty—great scientific discoveries can arise by accident, but great scientific theories rarely, if ever, occur in this fashion.

THE FATE OF THE UNIVERSE

Until the twentieth century, any discussion of the eventual fate of the Universe was conducted by philosophers and theologians, as the question could not even be accurately framed in a scientific setting. However, when Edwin

Hubble discovered that the galaxies were all receding from one another, this opened up the question to scientific debate.

If the galaxies are all flying away from one another, there are only three possibilities. The first is that the expansion remains unchecked, and that sooner or later each galaxy is alone in the Cosmos, unable to receive signals from any other galaxy. The second possibility is that there is enough mass in the Universe to reverse the expansion, and that the galaxies will all eventually collide with one another. The Universe, born in a big bang, will end in a big crunch. Third, there might be just enough matter to slow down the expansion to zero, but not enough to cause a big crunch. Scientists have calculated that the amount of matter needed to cause this last scenario is about three atoms of hydrogen for every cubic meter of space. This amount of matter is called the critical density.

If the only matter in the Universe is what we see through our telescopes, then the density of the Universe is only about 2 percent of the critical density, and the Universe would expand forever. However, there is a lot more matter out there that we cannot see through our telescopes.

The existence of this unseen matter, generally called "dark matter," was discovered by Vera Rubin, an astronomer who had earned her doctorate working for George Gamow, one of the authors of the big bang theory. Rubin decided to measure the speed at which various galaxies rotated. She observed that galaxies rotated much more rapidly than they would have if the only mass in these galaxies were the luminous mass we can see through the telescopes. The actual rotation of these galaxies could only be explained if there were large quantities of dark matter making them rotate faster.

Was there enough dark matter in the galaxies, and outside of them, to cause the Universe to contract in a big crunch? The answer to that question is still to be determined, and can only be answered by observation. However, a recent theoretical development suggests that the actual density of the Universe is the critical density.

Currently, the measured density of the Universe is about 10 percent of the critical density. At the time of the big bang, the actual density of the Universe had to be either the exact critical density, above it, or below it. Had the actual density been just a tiny bit above or below it, the 15 billion years or so of galactic expansion would have caused the current measured density to deviate tremendously from the critical density. Scientists were faced with a huge credibility problem: why was the initial density of the Universe exactly (to about sixty decimal places!) the critical density? This is known as the "flatness problem."

In 1980, Alan Guth proposed a revolutionary scenario known as "inflation." He theorized that for one unimaginably brief moment after the big bang, the Universe actually inflated (expanded) far more rapidly than it is doing now. No matter how the Universe had actually begun, this superfast expansion created conditions in which the measured density at this moment was extremely close to the critical density. The inflationary scenario also explained why the Universe looks pretty much the same in all directions. Although it is still a theory, it has a large following among cosmologists.

Another interesting question raised by Rubin's work is the nature of the dark matter: what exactly is it? At the moment, no one knows, but there are plenty of guesses. Some astronomers believe that dark matter consists mostly of run-of-the-mill objects such as stars that have died and no longer glow. Some think that neutrinos, which recently have been shown to have very small masses, actually have enough mass to close the Universe, because there are a tremendous number of neutrinos in the Universe. Finally, there is a contingent that believes that dark matter consists of particles we have not yet detected; these particles are described by the more exotic "supersymmetric" theories of the nature of matter.

When Vera Rubin graduated from Vassar in 1948 as the school's only astronomy major, she applied to Princeton for graduate study, only to be told that "Princeton does not accept women" in the astronomy program—a policy that was only discarded in 1975. Princeton is undoubtedly one of the great educational and research institutions of the world, but it was way behind the curve as far as civil rights are concerned—in the 1940s, it also had a policy of not allowing blacks to attend lectures.

Rubin eventually received her doctorate from Georgetown University working under the supervision of George Gamow, who was one of the first proponents of the big bang theory. Gamow wrote one of the first popular books on mathematics and science. A few years ago, I had the opportunity to re-read *One, Two, Three . . . Infinity*. Some of it is a little out of date, but it's still a great introduction to math and science.

THE DISCOVERY OF PLANETS OUTSIDE THE SOLAR SYSTEM

Just as the question of the origin of life on our own planet is one of the most important questions that science has yet to answer, so is the existence of life elsewhere in the Universe. Despite the legions of reports of alien abductions and visits by flying saucers, there is not a single confirmed shred of evidence pointing to the existence of life beyond the limits of the Earth's atmosphere.

Our Moon is totally dead. Life would have to be able to evolve and survive at a temperature of 1000°C in an atmosphere of carbon dioxide and sulfuric acid on Venus, and the Mariner landings on Mars gave hints of intriguing chemistry but no sign of biology.

Scientists feel reasonably sure that, in order for life to exist, it must evolve on a planet. However, since the detection of the planet Pluto by Clyde Tombaugh in 1930, no new planet has ever been observed—and Pluto has sadly been demoted from planetary status. The detection of Pluto was the culmination of a search lasting many years, and all attempts to find another planet in our own solar system have failed. To find a planet circling another star, and consequently thousands of times further away from us than Pluto, was until quite recently an impossible task.

Not only is a planet physically so small that it would be virtually impossible to see at so great a distance, it would also be hidden by the glare of the star it is orbiting. Despite the improvements in technology, for decades scientists felt that the optical detection of a planet would be virtually impossible. The only possibility, many felt, would be the reception of a signal from another life-form. Thus was born SETI, the Search for Extraterrestrial Intelligence.

However, there was another possible way to detect the existence of a planet. Although most people think that the planets in the solar system orbit the Sun, in reality the planets orbit the center of mass of the solar system, which is actually within the Sun. The Sun itself also orbits this center of mass, and this produces detectable wobbles in the Sun's rotation. Perhaps these wobbles could be detected in other stars.

The generally accepted theory of the evolution of the solar system is that the planets are the result of the gravitational collapse of a cloud of material circling the Sun. In 1984, astronomers Bradford Smith and Richard Terrile obtained photographs of a cloud of material circling the star Beta Pictoris. Even though no planets were found, this was felt to be an extremely hopeful sign, as it indicated that the processes that produced the solar system could occur elsewhere.

Meanwhile, other scientists were still trying to detect wobbles in other stars by using an extremely sensitive technique known as interferometry. In 1994, astronomer Alexander Wolszczan detected the first planets outside the solar system. Scientists initially had a hard time crediting Wolszczan's discovery, because the presumed planet was circling a pulsar. According to conventional theory, this planet should not have existed. A pulsar is created when a massive star explodes as a supernova, leaving a rapidly rotating neutron star. Why didn't the supernova explosion destroy the planet? As of now, no one knows—maybe the planet was captured from another nearby star, or maybe our theories of how pulsars are created are in error.

The Wolszczan planet could not possibly serve as a source for extraterrestrial life. For that, it is felt that a star much like the Sun would be the best bet. However, in the quarter-century since Wolszczan's discovery, the exoplanet business has boomed, thanks both to improved wobble detection and the addition of other weapons, such as transit detection, to the arsenal. There are now more than four thousand known exoplanets, including several that orbit sunlike stars in a zone that is habitable for life as we know it. Sometime in the next few years, we will actually have technology that will enable us to see the surface of some of these planets, and maybe detect signatures of life.

CHAPTER 2

The Earth

We have been as curious about the Earth, which lies under our feet, as we have been about the planets and stars that lie above our heads and have been out of reach for most of human history. Surprisingly, it has been almost as difficult to piece together an accurate picture of the Earth as it has been to piece together an accurate picture of the heavens.

If one looks at a sixteenth-century map of the surface of the Earth, it gets a lot right—especially with regard to Europe, Africa, and a large part of Asia. North America and South America are depicted less accurately, the Arctic still less accurately, and Australia and the Antarctic receive almost no mention. Although a reasonable amount was known about the surface of the Earth, almost nothing was known of Earth's history—how it came to be, how old it was, and how it was structured.

Five hundred years later, we have reliable answers to those questions. What we don't have is a reliable answer to what is going to happen to the Earth in the near term. We do know that some 2 billion years from now, the Sun will expand to swallow the Earth, but a much more pressing question is whether what we are currently doing will alter the habitability of the Earth in the next century or so.

The nearest exoplanet is Proxima Centauri b, about 4.2 light-years from Earth. Fortunately, it is located in the habitable zone, that region of space surrounding a star that conceivably could support human life. We may even be able to get a look at Proxima Centauri b sometime in the next few decades—but unless we actually develop some method of traveling through space at velocities close to the speed of light, it will take millennia—or longer—to get there.

Throughout human history, the Earth has been our home. It will be our home for the foreseeable future. It behooves us to understand how it functions and how we affect its functioning—because humanity isn't going anywhere soon.

Measurements

THE FIRST ACCURATE MEASUREMENT
OF THE SIZE OF THE EARTH

One of the chief activities common to all sciences is measuring.

Nowadays, geometry is not considered a science, but rather a branch of mathematics. However, the literal meaning of the word "geometry" is "the measurement of the Earth." Although several pre-Hellenic civilizations used different aspects of geometry, it was the Greeks who formalized the study of geometry, and used it both to examine their world and construct their civilization.

To the Greeks, the shape of the Earth was not in doubt. The sun and moon could both be seen to be circular, and during a lunar eclipse the shadow of the Earth, falling upon the moon, was also seen to be circular. Since the shadow was always circular no matter when the eclipse occurred, every shadow cast by the Earth must be a circle, and this was possible only if the Earth was a sphere. What was not known to the Greeks was the actual size of the sphere.

Eratosthenes was a man of far-reaching intellectual interests. He was a historian who made an attempt to establish an accurate chronology of all the dates since the Trojan War, and is recognized as the first man to realize the need for the accurate dating of events. He was a literary critic who wrote a treatise on Greek comedy. As a scholar he was so well-regarded that he was summoned from Greece to Alexandria to head the library at Alexandria, arguably the most important intellectual position of the day. He was the greatest geographer of the time, constructing a map of the known world that extended from the British Isles to Ceylon and from the Caspian Sea to Ethiopia. He was a mathematician who constructed a systematic way of determining prime numbers that is still known as the sieve of Eratosthenes.

However, the feat for which he is best known is measuring the size of the Earth. He realized that the cities of Alexandria and Syene lie on the same meridian, and made use of that fact to determine the circumference of the Earth. It was known that on a particular day of the year the Sun was directly overhead at midday in Alexandria. What Eratosthenes needed was someone to make a critical measurement, and he established the laudable precedent of paying for scientific research by hiring a man to travel the 800 or so miles south to Syene to measure the angle cast by a shadow at precisely the same time. Using this measurement, by simple geometry he was able to estimate the diameter of the Earth as approximately 25,000 miles, quite close to its actual value.

Eratosthenes also used this value to deduce geographical facts about the undiscovered portion of the Earth. He was able to estimate the size of the area of the known world from his experience as a geographer. Realizing that this formed a very small portion of the size of the Earth, he reasoned that the seas lying to the east and the seas lying to the west must actually form a part of a worldwide interconnected ocean. This deduction took nearly two thousand years to establish, but it was confirmed when Ferdinand Magellan circumnavigated the globe in the sixteenth century.

Eratosthenes was known to his colleagues as Beta, the second letter in the Greek alphabet, as he seemed to be the world's second-greatest authority on practically everything. When he was eighty years old, his sight failed. Depressed by his blindness, he committed suicide by voluntary starvation.

THE AGE OF THE EARTH

In an extraordinary work of biblical scholarship, James Ussher, an Irish bishop, culminated years of study with the announcement in 1650 that the Earth had been created at 9:00 a.m. on October 26 in the year 4004 BCE. The chronology determined by this date was printed in the margins of the King James Bible for two hundred years. However, more than one hundred years before Ussher completed his work, the German Georgius Agricola had noted the presence of mysterious bones in the Earth, and had coined the word "fossil" to describe them. In 1691, the English naturalist John Ray conjectured that these were the remains of extinct animals. This was a revolutionary point of view at that time, but later events were to prove it correct.

Considering the evident slowness with which living species change, the existence of fossils prompted one of two conclusions. Either the Earth was periodically swept by catastrophes that totally destroyed existing species, and then repopulated, or existing species evolved into others over long periods of time. The former view was obviously popular with those who felt that the Biblical view of creation was correct, while the latter appealed to a growing community of biologists and geologists. By 1830, the geologist Charles Lyell estimated that the oldest fossil-bearing rocks must be hundreds of millions of years old.

By the middle of the nineteenth century, other sciences were to become involved in this question. William Thomson (later to be knighted Lord Kelvin) was an infant prodigy. The son of a mathematics professor, he entered the University of Glasgow at the tender age of eleven, graduating second in his class. He was appointed to a professorship in 1846, at which time he decided to calculate the age of the Earth using the laws of cooling that were known from physics. Under the erroneous assumption that the Earth was originally part of the sun, he derived the result that the age of the Earth was approximately 100 million years.

This was not well received by the geologists, who were using other arguments to reach the conclusion that the Earth was considerably older. The astronomers and geologists engaged in heated debate for over half a century, but it took the science of chemistry to come up with an answer that satisfied everyone.

By the twentieth century, radioactive elements had been discovered, and had been shown to give off heat. This rendered Kelvin's argument invalid. In 1904, Bertram Boltwood was able to show that one radioactive element could decay into another, and that uranium would decay into lead, whereupon the decay would stop. By 1907 he had improved his results to the point where he could calculate the age of a rock by computing the ratio of uranium to lead, and concluded that minerals near his Connecticut home were more than 400 million years old. Thus was born the science of radioactive dating. Using this technique, rocks have been discovered that are 4.2 billion years old, and the current age of the Earth is estimated at 4.6 billion years.

The problem of determining the age of the Earth is obviously a compelling one, as religion, paleontology, biology, geology, astronomy, physics, and chemistry have all had a hand in its solution.

The name Kelvin is inextricably tied to heat, as it was a subject which fascinated him. Even though he was significantly in error in his use of heat to determine the age of the Earth, he was extremely accurate in arriving at one very important conclusion. Noting that a gas at 0°C lost 1/273 of its volume for every degree Celsius the temperature decreased, he reached the conclusion that it would have no volume at −273°C, and therefore it was impossible to achieve a colder temperature. He therefore suggested that a new temperature scale be adopted in which the coldest possible temperature was 0°. This temperature is now known as absolute zero, and measurements on this temperature scale are quoted in degrees Kelvin (°K).

Geology

THE THEORY OF UNIFORMITARIANISM

There is only one requirement for becoming a scientist, and that is the urge to satisfy a deep curiosity concerning the true nature of the world around us. However, there is no requirement on the preliminaries that must be fulfilled in order to become a scientist. James Hutton entered science after flirting with one career and embracing another.

It is rare that a person receives a medical degree but never practices or conducts medical research. However, after graduating from medical school, Hutton became an agricultural chemist. Sensing that there were financial opportunities in the budding chemical industry, Hutton established a factory for the manufacture of ammonium chloride. He did so well at this that he was able to retire at age forty-two to pursue his chief interest, the study of the geological structures of his native Scotland.

This was the period of the Industrial Revolution, and it was becoming clear that the study of geology had important economic consequences. Correct location of canals and railways depended on knowledge of geological conditions, to say nothing of the clues geology could yield concerning potential mineral deposits. When Hutton began his work, the world's most respected geologist was Abraham Werner, a German who believed that the Earth had originally been covered with water in which minerals had been dissolved. Over time, solids precipitated out of the water to form the various layers of rocks that covered the Earth. This theory came to be called neptunism, from the Greek god of the sea.

Werner was one of the first to call attention to the possibility that it had taken the Earth a long time to reach its current state. At the time of Hutton's investigations, conservative points of view once again dominated the intellectual landscape. As noted in the previous section, during the early seventeenth century, Bishop James Ussher had made a detailed study of the time periods in the Bible and had come to the conclusion that the Universe had been created in 4004 BCE. This date was unquestioningly accepted by many religious and secular authorities in Hutton's time, and to challenge it ran risks, although the sanctions imposed were not as drastic as those mandated by the Catholic Inquisition a century and a half earlier on Giordano Bruno and Galileo.

Through long study, Hutton became convinced of two fundamental ideas concerning geology. The first was that the processes that shaped the

Earth continued to operate even today, and they were processes that ran slowly and at a uniform rate. As a result, Hutton's theories were described by the word "uniformitarianism." Hutton was also convinced that the mechanism driving the changes was the internal heat of the Earth. As Werner's theory was called neptunism, Hutton's was called plutonism, in reference to the deity controlling the heated nether regions.

Hutton published his conclusions in 1785, in a book entitled *Theory of the Earth*. Although he reached several erroneous conclusions, among which was the idea that the Earth had neither beginning nor foreseeable end, many of his ideas form the basis of modern geology. Geological processes are currently seen as being driven by two different operating systems, the uniform ones described by Hutton, and the catastrophic ones such as meteor impacts, which are presently thought to be responsible for the extinction of the dinosaurs.

Hutton is universally regarded as the father of modern geology, but if the notes on which he was working had come to light earlier, he might have achieved even greater fame. In 1947, scholars examined a Hutton manuscript that had been undiscovered until then. In it, he outlined some of the basic ideas on evolution by natural selection that were to occur to both Charles Darwin and Alfred Wallace half a century after Hutton had suggested them.

THE STRUCTURE OF THE EARTH'S INTERIOR

As anyone who has ever experienced an earthquake knows—and there was one with an epicenter not ten miles from me not long ago—an earthquake is one of the most terrifying of natural phenomena. Because of the devastation earthquakes can cause, the effort to understand and predict them is one of the primary goals of geology. It is one of the fascinating detective stories in science that the effort to understand what causes the damage at the surface of the Earth has revealed what is going on inside the Earth.

Richard Oldham was an Irish geologist who, like many others, made an intensive study of earthquakes. In examining the records of earthquakes, Oldham showed that an earthquake generated two main types of waves. The first type of wave is called the P (for primary) wave. It is a compressional wave, alternately compressing and dilating the rock through which it travels. The other major type of wave is called the S (for secondary) wave. It is a shearing wave, and is responsible for the characteristic up-and-down shaking my wife and I experienced while sheltering under a doorway frame—you learn to do this if you live in California.

Besides the types of motion they induce, the P and S waves have other, different characteristics. P waves are much faster, and always are the first to arrive after an earthquake. A P wave can travel through both solid and liquid, but shearing is not possible in liquids, and so S waves only travel through solid rock. Two other important properties of waves are reflection and refraction—the refraction of light waves is responsible for the fact that a straw in a glass of water seems bent at the interface between water and air. Oldham showed that the different speeds of the P and S waves, combined with their reflective and refractive properties, could be used as diagnostic tools to probe the interior of the Earth.

Andrija Mohorovičić studied records from a Yugoslavian earthquake that produced a second set of waves that mirrored the first set. He concluded that the second set occurred when the first set bounced off a discontinuity that marked the dividing line between the surface of the Earth (known as the crust) and another distinct layer of material. This next layer is known as the mantle. The dividing line is known as the Mohorovičić discontinuity (or Moho).

It had long been suspected that at the center of the Earth there exists a solid metallic core, simply because the density of the Earth was known to be greater than the density of rock. Oldham was able to analyze the waves from numerous earthquakes to construct a simplified model of the Earth with a metallic core. In the next few decades, the sophistication of seismic detectors improved, and the database of earthquake records expanded significantly. These advances made it possible for Inge Lehmann, a Danish geologist, to demonstrate that the Earth's core actually consisted of two distinct layers, an outer liquid layer and an inner solid core. As a result of these efforts, the structure of the Earth is now known. Even though there are local variations, this consists basically of a crust anywhere from 10 to 40 kilometers thick, a mantle of molten rock that extends for another 2,800 kilometers, a liquid outer core some 2,200 kilometers thick, and a solid metallic inner core with a radius of 2,200 kilometers.

A truly great earthquake has the power to shake the entire Earth and set it "ringing" like a massive spherical bell. Subtle information can be obtained from the various different "tones" with which the Earth rings. The devices used for this analysis are tomographic scanners, similar in principle to the CAT scans used in hospitals. It was also discovered that the Sun also rings like a bell, and the techniques employed in analyzing the Earth are used in the new science of helioseismology to investigate the structure of the Sun.

THE THEORY OF PLATE TECTONICS

When an artist's life is tragic, such as in the case of Mozart or Van Gogh, it sometimes makes its way into the popular culture in the form of plays or movies. When a scientist's life is tragic, it does not seem to attract the same attention.

Alfred Wegener was a respected German meteorologist in the first two decades of the twentieth century. Wegener's interests extended beyond meteorology, and like others before him, he was intrigued by the apparent close fit between the west coast of Africa and the east coast of South America. Unlike the others, he did not confine his investigations to the shape of the two continents, but examined the geologic and fossil records of both continents. It appeared to Wegener that similar rock strata and fossils could be found on both continents. As a result, he proposed a theory of "continental drift." In this theory, all the continents had at one time formed a single land mass, which had later fractured into the various continents. Over hundreds of millions of years, the continents had drifted apart.

There was a major difficulty with this theory. At the time Wegener propounded it, no mechanism was known that would enable the continents to drift apart. As for the similar fossil records in Africa and South America, the geologists of the time proposed the existence of land bridges between the continents, which had subsequently sunk beneath the sea. Wegener, the geologists of the time suggested, should stick to meteorology.

Wegener had long been interested in Greenland, and had made three successful expeditions there. His fourth expedition was a mission of mercy, an attempt to bring food to a group of researchers who were running low on supplies. When he arrived, there were not enough supplies remaining to enable everyone to ride out the winter, so Wegener and a colleague took dog sleds to try to make it to another camp. They never made it.

Ten years later, the world was plunged into the Second World War, and it became important to obtain maps of the ocean floor. One person who worked on this problem was Harry Hess, an American geologist who actually attained the rank of rear admiral in the U.S. Naval Reserve. In the early 1960s, F. J. Vine and D. H. Matthews made a startling discovery concerning the structure of the ocean floor. It had been known since 1929 that the earth's magnetic field continually reversed its polarity after several hundred thousand years. Vine and Matthews discovered evidence that the direction of the Earth's magnetic field was recorded in the rocks on the ocean floor in adjacent parallel strips. The youngest strips are next to a sub-oceanic valley with mountains on either side, which is known as a rift valley. The further one travels from the rift, the older the magnetized strips of rock become.

Hess proposed that new ocean floor was formed by volcanic action at the rift, and that the creation of new floor wedged the old strips further apart. This meant that new ocean floor was being continually created, and since the Earth was not getting larger, the old surface must be destroyed. Hess's theory was that the older portion of the surface would be destroyed by being submerged and later melted by the intense heat of the Earth's interior.

This theory was refined during the 1960s, as it was discovered that the Earth's surface consists of approximately a dozen large plates, which are continually moving and colliding with one another. As the plates collide, mountains are formed and one plate is submerged under the other, with the destroyed plate surface being replaced by volcanic action at the rift valleys. This theory, known as plate tectonics, not only explained why earthquakes occurred at particular locations (where the plates were colliding), but also provided the mechanism for Wegener's continental drift—the continents drift on the moving plates. Like Van Gogh and Mozart, Wegener had been vindicated by a subsequent generation.

Although Hess was an acknowledged expert on the Earth's oceans, he also lent his expertise to NASA as it prepared for the first landing on the moon. Just as Wegener was not to see the results of his work, neither was Hess, who died a month before the successful mission of Apollo 11.

The Earth's Surface

DYNAMICS OF THE ATMOSPHERE AND THE OCEAN

One of the most important discoveries in the earth sciences has been the complicated interrelationship between the oceans and the atmosphere. It is a story of exploration, of data gathering, of mathematical analysis, and of physical modeling—and there's a lot of science that is essentially the same type of story.

All the oceans of the world are interlinked, but they are not just stagnant pools of water. In the middle of the eighteenth century, the Board of Customs in Boston complained that mail packet ships from England took two weeks longer to make the transatlantic crossing than did Rhode Island merchant ships. Benjamin Franklin asked a Nantucket sea captain if he could find an explanation, and the captain told him that the American ships avoided the Gulf Stream on the westward leg, but the British ships paid no attention to it. Since the Gulf Stream moved at three miles per hour relative to the surrounding water, the British ships were effectively trying to swim upstream.

Franklin was the first to draw a chart of the Gulf Stream, the greatest of the ocean currents. The explorer Alexander von Humboldt discovered another major ocean current off the coast of Peru. Now called the Humboldt Current, its behavior has a profound effect on world climate. At irregular intervals, the normally cold Humboldt Current is deflected away from the coast by a rush of warm water surging down from the equator. This is disastrous for the local economy, as the anchovy harvest on which it depends is greatly reduced by warm water. More importantly, this El Niño condition, as it is called, has a profound effect on the world's climate, often producing floods in the United States and drought in Africa.

The first great advance in understanding the physics of the ocean came from an analysis performed in 1835 by a French physicist, Gaspard de Coriolis. He showed that the rotation of the Earth deflected moving air and water eastward when moving away from the equator, and westward when moving toward it. This deflection, which was mathematically proven some twenty years later by William Ferrel, is known as the Coriolis force, and it has a strong influence on the formation of both wind and water vortices. In the northern hemisphere, these vortices circulate clockwise, while they circulate counterclockwise in the southern hemisphere, as can be seen when watching water swirl down the drain and rotating in different directions depending on whether you are north or south of the equator.

The twentieth century saw an intensive investigation of the relationship between the atmosphere and the ocean. Because of the importance of atmospheric and oceanic behavior on the Scandinavian countries, many of the major contributors in this area have been Scandinavians. Three generations of the Bjerknes family—Carl, Vilhelm, and Jacob—have devoted their lives to the study of the oceans and the atmosphere. The fronts that appear on the daily weather map you see on TV use symbology devised by Jacob Bjerknes.

Vilhelm, the second of the three generations, not only corrected errors in his father's analyses, but was an extremely inspirational teacher. Among his students were his son (naturally), and two of the twentieth century's foremost scientists in this area. Vagn Ekman studied the effect that the Coriolis force had on the top layer of the ocean; this layer is now known as the Ekman layer. Another student, Carl-Gustaf Rossby, is responsible for discovering the jet streams, those fast-moving currents of air which influence not only the daily weather, but also the time necessary for airplane travel. In a sense, the jet stream plays a role in the atmosphere similar to the one the Gulf Stream plays in the ocean—affecting not only weather but transportation.

One of Rossby's contributions to meteorology was that he was among the first to apply computers to the problem of weather forecasting. Computerized weather forecasting has improved substantially over the last half-century; the seven-day forecasts of today are as accurate as the two-day forecasts of the early 1970s. Computers are also currently being used to analyze and forecast future climate changes. A disturbing discovery is that the coupled interplay of oceanic and atmospheric currents has resulted in bizarre climatic shifts in the past, with the world being suddenly jolted into either glacial or tropical conditions in extremely short periods of time. The next great climatic shift may not come from the greenhouse effect, but from conditions deep below the surface of the ocean. And warming and cooling seem to be interlinked. Recent studies have shown that there is a strong correlation between the warming of the Arctic and the severe winters in North America, and the cause of this is the shifting of a current deep in the Atlantic Ocean.

THE ICE AGES

Occasionally a great scientific discovery is not so much a brand-new idea as it is a rediscovery of a previously existing one. For example, in medieval Europe it was common practice to cover wounds with moldy bread; centuries later, Alexander Fleming would discover the antibiotic powers of the penicillin mold.

Perhaps it is appropriate that Louis Agassiz, the scientist who first made the ice age a credible scientific theory, was Swiss, because the icy glaciers of Switzerland were in large part responsible for this theory. However, the idea of an ice age did not occur first to the scientific community, but rather to the mountaineers of Switzerland, who were very familiar with glaciers. In 1815, the mountaineer Jean-Pierre Perraudin observed scars on hard rocks that did not weather at the base of a valley. Having seen similar scars on rocks near glaciers at the top, he wrote that the only explanation was that glaciers had once filled the entire valley. He communicated this to a well-known naturalist, who was impressed but not completely convinced.

In 1815, Louis Agassiz was eight years old. After graduating from college with a degree in zoology, he took a position as a professor of natural history at Neuchâtel, Switzerland, where he produced a well-received five-volume series on fossil fishes. Agassiz was a naturalist who loved exploring the countryside, and these explorations brought him into contact with large boulders in the middle of valleys, which obviously had been transported to their present locations. From several scientists and non-scientists he heard the conjecture that the boulders had been brought from high atop the mountains to the valleys by means of glaciers that had long since receded. Agassiz decided to investigate for himself.

The alternative source of movement for these boulders was the "Great Flood" described in the Bible. Raging rivers were known to have the ability to move large boulders, and no one had yet amassed evidence to demonstrate that glaciers could also do the job. Agassiz found that glaciers generally terminated in rocks and boulders, and these rocks bore scratches and grooves similar to those found by Perraudin. In 1839, Agassiz found a cabin that had been built on a glacier in 1827, and had moved a mile down the glacier from the original site. He then pounded a straight line of stakes deep into the ice; within two years they had not only moved but were now shaped like a U. This showed that the ice in the center moved faster than the ice on the sides, which was slowed by friction with the surrounding mountains.

Agassiz was now persuaded that there had indeed been an ice age. He found evidence that glaciers once existed in the British Isles. As a result of his years of research, the existence of an ice age was finally established.

Modern science has uncovered the fact that ice ages have been a recurring phenomenon; evidence exists that there were ice ages hundreds of millions of years ago. Now that we know that ice ages are a part of our history, the question arises: what causes them? One of the most ambitious efforts in this direction is the theory of the Yugoslavian physicist Milutin Milankovitch, who published three papers between 1912 and 1914. In these he hypothesized that there were two astronomical cycles that played a major role in determining Earth's climate—the 41,000-year cycle of the inclination of the Earth's axis, and the 22,000-year oscillation of the Earth-Sun distance. Milankovitch's theory has generated a good deal of interest in the meteorological community. Even though his original formulation seems to have been disproved, every so often a scientist finds another cycle that, when combined with the ones cited by Milankovitch, does an increasingly good job of predicting climate conditions.

Another fascinating question is: when is the next ice age? In an era when we are continually reminded of the possibility of overheating due to the greenhouse effect, it would be ironic if we were next to suffer through a trial by ice rather than a trial by fire. Recent computer simulations have raised the disturbing possibility that the meandering of deep, cold, ocean currents is chaotic in nature, and a dislocation of these currents could flip the Earth into an ice age in less than a century.

THE EFFECT OF MAN ON THE EARTH

It is only within the past few decades that ecology, a word coined by the German philosopher and biologist Ernst Haeckel to describe the interrelationship between living things and the environment, has become a recognized scientific subject. However, it was the Swedish chemist Svante Arrhenius who was responsible for the first discovery concerning the possible effects of the activities of man on the environment.

Arrhenius was a top-notch chemist whose theory of the behavior of ions resulted in his winning one of the first Nobel Prizes. In 1896, Arrhenius noted that the gas carbon dioxide had the property of allowing the high-frequency sunlight that the Earth received by day to pass through it, but it reflected the low-frequency infrared light that the Earth reradiated as heat by night. This meant that a buildup of carbon dioxide in the atmosphere, which Arrhenius noted was taking place because of increased industrialization, could be accompanied by an increase in heat. This was the first discussion of the greenhouse effect. Later in the twentieth century it would be shown that the planet Venus had succumbed to a runaway version of the greenhouse effect, resulting in an atmosphere of carbon dioxide at crushing pressures and a planetary temperature of 475°C.

Ecology as a science was neglected during the first half of the twentieth century. The key development in its emergence occurred as the result of a correspondence between a naturalist author and a friend who owned a bird sanctuary. Rachel Carson was an aquatic biologist with the U.S. Bureau of Fisheries who had abandoned her youthful interest in writing to major in zoology in college. While with the bureau, she prepared a series of radio broadcasts on underwater life, and eventually published her first book, *Under the Sea*, in 1941. Her second book, *The Sea Around Us*, published ten years later, was an instant classic and a huge financial success.

The financial freedom she achieved enabled her to observe and write about nature. A friend of Carson who owned a private bird sanctuary wrote to Carson to describe the appalling effects of DDT spraying on the birds within the sanctuary. DDT was an insecticide that had proved tremendously beneficial in wiping out mosquito populations and greatly reducing the occurrence of malaria, but its effect on wildlife had not been thoroughly investigated. Carson's study of the problem resulted in the book *Silent Spring*, unarguably the single most influential book ever written on environmental problems. Published in 1962, by the end of the year legislators had introduced over 40 bills concerning pesticide regulation, and the environmental movement as we know it today was born.

Carson died in 1967, not living long enough to see the banning of DDT in the United States in 1972. Environmental studies are now part of the curriculum of many colleges and universities. A heightened awareness of the impact of man on the environment has caused a major change in the actions of business and industry, which must now file environmental impact reports before major construction projects are authorized.

A measure of the increased consciousness concerning environmental problems can be seen by looking at the history of chlorofluorocarbons (CFCs), chemicals commonly used in refrigeration. In 1974, Mario Molina and F. Sherwood Rowland warned that CFCs may be contributing to the destruction of the ozone layer that protects the Earth from ultraviolet radiation. This report was at first taken lightly. In the 1980s, satellites detected the growth of a hole in the ozone layer over the Antarctic. As a result, the Montreal Protocol resulted in the decision to totally phase out CFCs by 2000. In 1995, Molina and Rowland received the Nobel Prize in chemistry for their work.

THE DISCOVERY OF CHAOTIC PHENOMENA

We've discussed how important measurement is to the sciences. In fact, it has long been realized that numbers so completely describe a large class of phenomena that if there were a way to predict the numbers, one would in essence be predicting the phenomena. As a result, the development of affordable electronic computers during the late 1950s was welcomed by a host of scientists eager to use the devices to construct numerical models with which they could simulate processes in the real world.

In late 1961, Edward Lorenz, a meteorologist at the Massachusetts Institute of Technology, had programmed one such computer to simulate the weather. The movement of the atmosphere was an example of fluid flow, the equations for which were well known but impossible to solve directly. With a computer, it was possible to generate numerical approximations to the exact solutions of these equations. Lorenz developed a computer model that would print numbers characterizing a single day's weather every minute. He became accustomed to some of the numerical patterns that the computer would generate, much as one becomes accustomed to the patterns of the weather, which are familiar but never exactly predictable.

In order to start the model, it was necessary to give it initial conditions, an assumed state of the weather from which the model could begin calculating. One day Lorenz wanted to examine a particular sequence of days more closely, and so he took a shortcut. Rather than start all over again, he took the output from the computer for a day midway in the sequence, and used that output as the initial conditions. For a while, the computer results duplicated the previous run, just as one would expect. Then, very slowly, discrepancies began to appear. After some time, the results of the second run bore no relation to the results of the first run.

It took Lorenz some time to realize what had happened. The computer had computed and stored data to six-digit accuracy, such as .318297, but typed them out only to three-digit accuracy, in this case .318. When Lorenz re-entered the numbers, he had entered only three digits. As a result, the computer was presented with an initial condition of .318, rather than .318297. This minuscule difference in initial conditions would have major effects later in the computer run.

The fact that minuscule initial differences can have subsequent profound consequences is now known as the "butterfly effect." The name comes from the notion that whether or not a butterfly flaps its wings in Hawaii can determine whether or not there is a tornado in Kansas three weeks later.

The butterfly effect was to be the first example of a host of processes that are now known as chaotic phenomena. Prior to their discovery, a process would be described as either deterministic or random. A simple example of deterministic phenomena would be the orbits of the planets, which are so reliable that it is possible to predict eclipses centuries in advance. A flip of a coin, or the decay of a radioactive atom, is an example of a random process. Chaos is the study of those phenomena that appear on the surface to be predictable, but whose predictability turns out to be intrinsically limited.

Sometimes a discovery in science triggers a re-examination of many other areas. Just as Newton's mechanics spurred the search for mathematical laws in areas encompassing all branches of human knowledge, Lorenz's work has led to the discovery of chaos in areas as diverse as the fibrillation of the heart during a heart attack, and the behavior of stock markets during financial crashes. The realization that some phenomena are chaotic has expanded the way we describe the Universe.

It has also exacerbated the fear that at some stage, we may inadvertently load on the straw that breaks the camel's back of the climate system. We do not know whether what we are doing will precipitate an ice age, or a runaway greenhouse effect like the one on Venus. But we had better pay attention to the warnings the simulations give us, as one day we may awake to the fact that those simulations have indeed been the distant early warning (DEW) line for an ominous reality.

CHAPTER 3

Chemistry

Complicated stuff is often made up of simple stuff. A lot of science is organized around three important questions: what are the basic building blocks, how are they arranged to form more complicated structures, and what can we humans do to form these more complicated structures.

When I was growing up, one of the major chemical companies sponsored a TV show I enjoyed watching. The motto of the company was "Better Living Through Chemistry." And arguably, of all the sciences, chemistry may be able to claim the prize for most contributions to making life better.

Yes, the life sciences are tremendously helpful when we're sick—but thankfully, most of the time we aren't sick. And it's hard to imagine life without some form of energy harnessed through our knowledge of physics—life wouldn't be the same without universal electrification—but you can run a reasonably complicated and moderately advanced society without it.

But we've been taking advantage of chemistry ever since we first used fire (arguably the first chemical reaction to be successfully harnessed), and since we first used a chemical reaction (even more arguably fermentation, for the purpose of creating alcoholic beverages). And every day you use a host of chemical products that have made your life demonstrably better. "Better Living Through Chemistry" could easily serve as a one-sentence description of the success of Western civilization. It is hard to pass an hour in any day that is totally unaffected by the consequences of our knowledge of chemistry.

As supporting evidence for the fundamental role played by chemistry, there's a branch of chemistry in practically every other science—from astronomy to zoology, and I don't believe there's a whole lot of astronomy in zoology, or vice-versa. But chemistry permeates every branch of science because chemistry is the quintessential subject that investigates stuff—how it is made, how to make it, and how to make new stuff. And that includes stuff that, as far as we know, exists only here on Earth, because it was nowhere to be found on Earth until we made it, and we have as yet received no reports of stuff-making aliens. At least, no reliable reports.

Organizing Principles

Isaac Newton, my candidate for the most influential scientist in history, is best known for his contributions to math and physics. However, he dedicated a number of years in an attempt to do for alchemy what he had done for physics, and didn't accomplish anything of note.

In devising his theories of mechanics and gravitation, Newton had a lot to work with. Tycho Brahe and Johannes Kepler had provided valuable data on the orbits of the planets, and Galileo had come up with his law of falling bodies. But there existed no data on subjects related to chemistry that would have helped him, as the rules that formed the basic organizing principles of chemistry were not to be discovered until nearly a century after Newton had done his seminal work in physics.

THE LAW OF CONSERVATION OF MASS

One of the most difficult things to accomplish in science is to overturn a widely supported theory. The picture that science is a smoothly progressing march to the truth is almost completely wrong. Many of the great advances in science have had to overcome well-established, but erroneous, opposing views.

A case in point is the phlogiston theory of combustion, which was the dominant viewpoint on this subject for more than a century. Centuries before Christ, the Greeks had propounded the theory that everything in the Universe was constructed of four elements: earth, air, fire, and water. The pursuit of knowledge, reborn after the Dark Ages, needed a foundation on which to build. That foundation was often the theories of the ancient Greeks.

One of the great mysteries of the seventeenth century was the nature of fire. The German chemist, Georg Stahl, building on both Greek theories and the experimental knowledge that was gradually beginning to be developed, expounded what was known as the phlogiston theory. This theory held that all combustible materials contained a substance called phlogiston. Upon burning, the substance released its phlogiston into the air. Substances that burned especially well were rich in phlogiston; substances that did not burn contained no phlogiston.

On the surface, this was certainly a plausible theory. It explained the fact that some substances burned better than others, and it also explained why, after burning, a substance was no longer capable of combustion. However, all attempts to isolate phlogiston met with defeat.

Not only that, there were experiments whose results were at odds with the phlogiston theory. When mercury or tin was burned, the weight of the resultant material after burning was greater than the weight of the material before burning. Phlogiston theory predicted that the substances, having lost their phlogiston, would be lighter after burning.

Faced with the results of these experiments, the proponents of phlogiston theory did what many scientists since have done—attempted to modify the theory to fit the observed data. Phlogiston was hypothesized to be a substance without weight, or possibly even lighter, and so under the right circumstances, a substance could gain weight by losing phlogiston.

By the latter portion of the eighteenth century, air had been shown to be a mixture of various gases, one of which (oxygen) had been shown not only to enhance combustion, but to be necessary for respiration. Enter Antoine Lavoisier, a French lawyer and tax collector turned chemist. He performed an experiment that was both simple and elegant. First he heated mercury in the presence of oxygen, carefully weighing the amount of oxygen and mercury before, and the amount of material (mercuric oxide) and oxygen after. Then he heated the mercuric oxide to the point where the oxygen was released, again carefully measuring the amount of material with which he started and with which he finished. From the results of this experiment, he was able to conclude not only that combustion consisted of combining with oxygen, but that in a chemical reaction—even though the substances involve may change—the total quantity of reactants doesn't change. With one experiment, Lavoisier had not only demolished the phlogiston theory, but established one of the most fundamental laws of chemistry, that of conservation of mass.

The idea of conservation, although not in the quantitative form in which Lavoisier stated it, was broached more than two millennia before by the Greek philosopher Epicurus, who believed that "the totality of things was always such as it is now, and always will be." The Arab scientist Nasir-al-Din-al-Tusi nailed it more precisely when he wrote, "A body of matter cannot disappear completely. It only changes its form, condition, composition, color and other properties and turns into a different complex or elementary matter." But it was Lavoisier who actually established it quantitatively.

Lavoisier had the good fortune to be married to a woman who was of substantial help to his work, but had the bad fortune to be living in France during the time of the French Revolution. His work as a tax collector made him a natural target of the political frenzy that was sweeping the nation, and he was condemned to death by guillotine. When it was argued that Lavoisier was a great scientist, the reply from the presiding judge was, "The Republic has no need of scientists." Sadly, the last few years have seen substantial evidence that we are living in an era—and a republic—in which many have a similar view as the presiding judge.

Commenting later on his death, the brilliant mathematician and physicist Joseph-Louis Lagrange declared, "It took but a moment to sever that head, though a hundred years perhaps will be unable to replace it."

THE ATOMIC THEORY

In 1961, the brilliant physicist Richard Feynman began the basic physics course at Caltech with the following words: "If, in some cataclysm, all of scientific knowledge were to be destroyed, and only one sentence passed on to the next generation of creatures, what statement would contain the most information in the fewest words? I believe it is the atomic hypothesis . . . that all things are made of atoms—little particles that move around in perpetual motion."

The idea that all things are made of atoms, the smallest particles in the Universe to retain their identity, goes back to the Greek philosophers. But it is one thing to speculate on the ultimate constituents of matter, and quite another thing to develop a workable theory that not only explained, but predicted. At the outset of the nineteenth century, it was known that substances such as hydrogen and oxygen were elements, and that elements were capable of combining into compounds. Water, for instance, was a substance made by combining hydrogen and oxygen, and the same quantities of hydrogen and oxygen always produced the identical quantity of water. What mechanism could account for this?

John Dalton, a Quaker schoolteacher living in Manchester, England, spent the summer of 1803 pursuing an extension of the Greek theory of atoms. Unlike the Greeks, whose atoms were philosophical constructs, Dalton's atoms possessed a tangible physical property: weight. As Dalton wrote: "An enquiry into the relative weights of the ultimate particles is, as far as I know, entirely new. I have lately been prosecuting this enquiry with remarkable success." Dalton realized that, if one were to hypothesize that each element consisted of identical atoms, all having exactly the same weight, this would account for the manner in which the elements combined to produce compounds.

Science has always been a highly conservative and hardheaded endeavor, and new ideas tend to be treated with reserve bordering on skepticism. However, so brilliant was Dalton's atomic theory, and so strong its predictive power, that it was accepted virtually instantaneously. During the course of the next hundred and fifty years, the physical properties of atoms were determined with ever-increasing accuracy, even though it wasn't until the 1980s that atoms were first actually seen. Without Dalton's atomic theory, chemistry would be reduced to the hit-and-miss hodgepodge of mixing and heating that characterized its predecessor, alchemy, and we would all be significantly the poorer for it.

Dalton was a man of exceptionally regular habits. Every Thursday he would take a walk through the English countryside to play bowls (no bowling alleys existed in the nineteenth century), and every day for almost sixty years he would meticulously record the temperature, rainfall, and air pressure. During a lifetime he recorded more than 200,000 meteorological measurements, a database probably unparalleled for the era, and one that was never put to any noteworthy use. Or was it? Perhaps a lifetime of accumulating and reflecting upon meteorological data helped create the mindset that enabled Dalton to devise the atomic theory.

AVOGADRO'S HYPOTHESIS

The list of great scientists includes many individuals who did not start out to be scientists, but who embarked initially upon another career, such as law (one suspects that the list of great lawyers who started out as scientists is an extremely short one). A case in point is Amadeo Avogadro, who received a doctorate in law, practiced for three years, and then became a scientist.

This transformation took place roughly around the turn of the nineteenth century, when much of the scientific world was turning its attention to Dalton's atomic theory. Chemistry had burgeoned during the latter portion of the eighteenth century, and there was a concerted effort to explain every experimental result in terms of the atomic theory.

One such result was Gay-Lussac's law of combining volumes. A careful experimenter, Joseph Louis Gay-Lussac had discovered that if two gases were combined at equal temperature and pressure, they combined in simple whole-number ratios. When hydrogen and oxygen combine to produce water, 2 volumes of hydrogen will combine with 1 volume of oxygen, producing 2 volumes of water vapor.

From this information, Avogadro reached a startling but logical conclusion: at the same temperature and pressure, equal volumes of gases must contain equal numbers of particles. This result is known as Avogadro's hypothesis.

His reasoning was simple and straightforward. On combining 2 volumes of hydrogen with 1 volume of oxygen, there is no hydrogen or oxygen left uncombined. Under the assumption of Avogadro's hypothesis, 2 hydrogen particles would have combined with 1 oxygen particle to produce 2 water vapor particles, and no hydrogen or oxygen would be left uncombined.

Avogadro's hypothesis produced an immediate useful dividend. By that time, it was known that the chemical formula for water was H_2O. Using Dalton's atomic hypothesis, 2 molecules (a word coined by Avogadro) of water required 4 atoms of hydrogen (supplied by the 2 hydrogen particles) and 2 atoms of oxygen (supplied by the 1 oxygen particle). Therefore, each particle (molecule) of hydrogen must consist of 2 atoms of hydrogen, and each particle of oxygen must consist of 2 atoms of oxygen.

It later proved possible to extend Avogadro's hypothesis to all substances. Equal atomic weights of substances have the same number of particles. The atomic weight of carbon is 12, and the atomic weight of a molecule of ammonia (NH_3) is 17, so 12 grams of carbon contain the same number of particles as 17 grams of ammonia. An atomic weight of any substance contains "Avogadro's number" of molecules of that substance, a number that chemists simply call N. Later experiments would show that N is approximately 6 followed by 23 zeros.

Avogadro's hypothesis was unrecognized in its time, and it wasn't until the first International Chemical Congress in 1860 that it was fully appreciated. The Italian chemist Stanislao Cannizzaro showed that Avogadro's hypothesis could be used to compute the molecular weights of gases, and also represented a way to clear up the confusion that existed concerning the difference between atoms and molecules. Unfortunately, Avogadro did not live to see his ideas vindicated, having died two years earlier.

Avogadro's hypothesis actually occurred to John Dalton, who rejected it on intelligent but erroneous grounds. Dalton felt that the molecules of a gas were in close contact with one another, and if the gas molecules had different weights, it would be impossible for equal volumes to have the same number of particles. This reasoning applies perfectly to solids and liquids, but in gases the molecules are separated by substantial distances (substantial, that is, relative to the size of the molecules themselves). So convinced by this reasoning was Dalton that he suggested Gay-Lussac's experiments must be in error, and that Gay-Lussac should redo them.

THE PERIODIC TABLE OF THE ELEMENTS

Should you decide to cut up some beef, potatoes, carrots, and onions for dinner, and cook them together, you know precisely what you will get—beef stew. Moreover, you probably have a pretty fair idea of how it will taste. The situation was nowhere as simple for chemists in the middle of the nineteenth century.

By that time, the world's chemists had discovered sixty-three elements, the basic ingredients in the cosmic cookbook. The rules of the cosmic cookbook, however, remained maddeningly elusive. For example, when sodium, a lightweight fizzy metal, was "cooked" (chemically combined) with chlorine, a poisonous yellow-green gas, the result was common table salt, sodium chloride, a compound that was neither metallic nor gassy, poisonous nor fizzy. Until the rules of the cosmic cookbook could be discovered, the potential of chemistry would be limited to hit-or-miss activity.

One of the fundamental discoveries of science is that many phenomena in the natural world can be organized into a pattern. Dmitri Mendeleev, a Russian chemist, decided to try to organize the known elements into a pattern. To do so, he first arranged these elements in increasing order of atomic weight, the same physical property that had attracted the attention of John Dalton when he devised the atomic theory. He then imposed another level of order by grouping the elements according to secondary properties such as metallicity and chemical reactivity—the ease with which elements combined with other elements.

The result of Mendeleev's deliberations was the periodic table of the elements, a tabular arrangement of the elements in both rows and columns. In essence, each column was characterized by a specific chemical property such as alkali metal or chemically nonreactive gas. The atomic weights increased from left to right in each row, and from top to bottom in each column.

When Mendeleev began his work, not all the elements were known. As a result, there were occasional gaps in the periodic table—places where Mendeleev would have expected an element with a particular atomic weight and chemical properties to be, but no such element was known to exist. With supreme confidence, Mendeleev predicted the future discovery of three such elements, giving their approximate atomic weights and chemical properties even before their existence could be substantiated. His most famous prediction involved an element that Mendeleev called eka-silicon. Located between silicon and tin in one of his columns, Mendeleev predicted that it would be a metal with properties resembling those of silicon and tin. Further, he made several quantifiable predictions: its weight would be 5.5 times heavier than water, its oxide would be 4.7 times heavier than water, and so on. When eka-silicon (later called germanium) was discovered some twenty years later, Mendeleev's predictions were right on the money.

The periodic table has tremendous practical importance. If a substance is useful but has undesirable properties, it may be possible to modify it by substituting an element with similar properties. For those who must regulate their sodium intake, an acceptable alternative is "light" table salt—potassium chloride, made by substituting potassium, which lies directly under sodium in the periodic table, for the sodium in salt.

Scientists often develop their theories in surprising fashion. It was necessary for Mendeleev to engage in countless restructurings of his periodic table, as he had no idea at the start how many rows and columns would be required. To write down the results of each trial would tax anyone's patience. So Mendeleev constructed a deck of cards in which each card contained the name and properties of a specific element. Playing solitaire with this deck of cards made it easier and more entertaining to try the different possibilities for the periodic table. Incidentally, the nineteenth-century name for a version of solitaire was Patience, something that Mendeleev undoubtedly possessed in quantity.

Analysis and Synthesis

Analysis is the process of breaking apart; synthesis is the process of putting together. A typical chemical reaction features both. A simple example takes place when sodium hydroxide is combined with hydrochloric acid. Each compound breaks apart into parts, and these parts recombine into common table salt and water.

What distinguishes a chemical reaction is that different compounds emerge from the ones that were initially there. You can mix salt with water, and the water will become salty, but you've still got salt and water. The study of how compounds break apart, and how they are put together to create different substances, continues a process that began centuries ago and still enriches our quality of life. Not that it isn't without risk—many compounds have been created that have had a deleterious effect. But overall, we have produced, are producing, and will continue to produce better living through chemistry.

Electrochemistry

Humphry Davy may have been the Horatio Alger of science, a story of rags to riches. He was born burdened with debt, and as he did not enjoy school he dropped out at the age of seventeen to become a pharmacist's apprentice. As so often happens, the value of an education became apparent only after one leaves school, and so Davy began an extensive course of self-education.

His interests were initially diverse, as are the interests of many seventeen-year-olds, but they became focused after he read a book on chemistry by the great French scientist Antoine Lavoisier. At that time there wasn't much of a theoretical basis for chemistry, which probably bothered Davy not at all, as he was the quintessential experimenter.

His favorite guinea pig was himself. After he became the superintendent of an institution for the study of therapeutic uses of various gases, Davy would think nothing of inhaling the products of his experiment to test their effect. Fortunately, he never inhaled any cyanide, but he nearly suffocated twice, once when he tried to breathe hydrogen, and once when he tried to breathe carbon dioxide. However, one of these experiments paid very large dividends. Davy discovered that the gas nitrous oxide made him feel giddy and intoxicated, and also reduced the sensation of physical pain. His observations on this subject were initially ignored, but decades later nitrous oxide became the first chemical anesthetic. It is still used today.

Davy was probably the first scientist to popularize science. When he was hired to lecture for the Royal Institute, his lectures and demonstrations were so interesting and well-presented that he soon became a darling of London's high society. As an experimenter, Davy was brilliant rather than meticulous. He would get interested in a topic, and then experiment until boredom set in, after which he would switch to another topic.

After learning of Volta's development of the electric battery, Davy built powerful batteries, and in 1805 developed arc lighting, in which a strong current forces an electric arc to bridge the gap between electrodes. Like his discovery of the anesthetic powers of nitrous oxide, arc lighting had to wait decades before a practical use was discovered.

However, Davy's most noteworthy contributions were in the field of electrochemistry. It had been discovered that an electric current could be used to break up water into hydrogen and oxygen. At the time, substances such as potash, lime, and magnesia were suspected of being metallic compounds, but no one was able to demonstrate this. Davy used his powerful batteries to pass an electric current through molten potash. This liberated globules of the as-yet-undiscovered element potassium. On seeing the appearance of the shiny metallic drops, Davy danced around his laboratory in glee. Within a week, he had isolated metallic sodium from soda (it is easy to see where Davy found the names for his discoveries). Soon he had discovered the elements magnesium, barium, strontium, and calcium as well.

Davy's experiments showed the importance of electrochemistry to the scientific world. However, perhaps Davy's greatest contribution to science was his hiring of Michael Faraday as his assistant. Where Davy was impetuous, Faraday was meticulous. Davy discovered elements, but Faraday discovered laws. Among these were Faraday's laws of electrolysis, which demonstrated that there is a quantitative relationship between electricity and chemistry. These laws would later prove to be very important in demonstrating that electricity was a stream of particles, a discovery that initiated atomic physics.

Even as Davy was receiving the plaudits of the scientific world, he could sense that his assistant Faraday would eventually supersede him. Davy became jealous of the acclaim that Faraday was receiving, and when Faraday was nominated for membership in the Royal Society, there was only one negative vote—Davy's. It is conceivable that this action was the result of the systemic poisoning Davy had suffered in his early days as an experimenter, when he would inhale or taste any compound. His health began to deteriorate substantially when he was only in his thirties. He suffered a stroke at the relatively young age of forty-nine, and died two years later.

THE DYE MAUVE AND THE BIRTH
OF SYNTHETIC ORGANIC CHEMISTRY

We live in a world of synthetics. Many of the products we use every day are made of substances that man has created through techniques that began approximately 150 years ago, when the Royal College of Science in London decided it needed a good course in chemistry. At the time, the best chemists were Germans, but fortunately for the Royal College, they had a German connection. Prince Albert, Queen Victoria's husband, was German, and he suggested a young German named August von Hofmann, who happily accepted the job.

Hofmann had numerous research interests, including the study of coal tar, an unattractive gummy black residue given off by coal when it is burnt. Coal tar had been shown to be a source of several useful organic compounds, including benzene and aniline, which Hofmann had managed to obtain from coal tar.

One of Hofmann's best students was a bright, industrious seventeen-year-old named William Perkin. Hofmann offered Perkin a job as an assistant, which Perkin eagerly accepted. Perkin was so enthusiastic that he built his own laboratory at home, in order to pursue his research when not at the Royal College.

At the time, Britain was building an empire in the Far East, and many of its best and brightest were succumbing to malaria, which was an ever-present threat. It was known that quinine, found in the bark of the cinchona tree, would cure malaria, but the demand for quinine greatly exceeded the supply. Chemists had deciphered the molecular composition of quinine, and Hofmann suggested that Perkin try to synthesize it from coal tar.

This was not a complete shot in the dark. One of the coal tar compounds had roughly half of the atoms of each element needed for quinine, so joining two of the molecules together and adding some missing atoms might produce quinine. With the aid of hindsight, this approach had no chance because even though the composition of quinine was known, the molecular structure of quinine would have precluded such an approach.

One day, after a fruitless attempt to synthesize quinine, Perkin ended with a blackish goo. He decided to add some alcohol, and the concoction turned a beautiful shade of purple. Perkin was instantly aware that this accident might have serendipitously produced a useful and valuable compound.

At the time, all existing dyes were natural, and some were extremely expensive. Purple had always been a rare and much-admired color. In ancient Rome, the only shade of purple came from shellfish and only the nobility could afford it (from which comes the expression, "born to the purple").

Opportunity may have knocked only once, but Perkin recognized the sound. He obtained a patent, went into business, and within six months was producing a dye he called "aniline purple," which was superior to any other dye in its color range. In a marketing triumph, the name of the color was changed to "mauve," and Perkin soon found himself both rich and the world's leading authority on synthetic dyes.

The dam had been broken. Within years, virtually all natural dyes had been synthesized, and cheap and easily obtainable colored fabric began to brighten the world. The techniques of synthesis obviously could be used on other organic compounds. By the twentieth century, the game was not only to synthesize natural organic compounds, but create new ones never seen in nature. The plastics industry that has so revolutionized our lives is unquestionably the result of Perkin's accidental discovery.

Perkin clearly possessed the Midas touch when it came to chemistry. After ten years or so in the dye business, he was wealthy enough to retire and pursue his first love, research. One of his first achievements was the synthesis of coumarin, which is responsible for the pleasant smell of new-mown hay. This represented the beginning of the synthetic perfume industry that, like synthetic dyes, has annual revenues in the billions of dollars.

CHEMICAL BONDS

By the middle of the nineteenth century, chemists had confirmed the correctness of Dalton's theory that every compound was constructed using a fixed ratio of elements. Water was written H_2O (as it still is), but the mechanism by which two atoms of hydrogen combined with one atom of oxygen to produce water was a mystery.

The first person to make a noticeable dent in this problem was the German chemist Friedrich Kekulé. At the time, chemists had formulated a rough idea of the valence, or combining power, of each element. Kekulé came up with the idea that each molecule had a structural pattern, which could be represented by lines joining atoms. For example, ethyl alcohol, which had the chemical formula C_2H_6O, could be shown as

H H

| |

H - C - C - O - H

| |

H H

In addition to providing the first hint of pattern behind the structure of a molecule, Kekulé's notation enabled chemists to realize how two different isomers, which were different chemical compounds with the same ratio of elements, could exist. Isomers would simply be different structural patterns using the same ratio of elements.

One of Kekulé's greatest triumphs came in unraveling the structure of benzene, which was known to have the chemical composition C_6H_6. Try as he might, Kekulé was unable to fit benzene into his scheme until one day, half-awake and half-asleep, he dreamed of a snake rolling down a hill with its tail in its mouth. Thus was born the benzene ring, in which the six carbon atoms are linked in a hexagon by alternating single and double bonds, with a hydrogen atom sticking onto each carbon atom like a spoke protruding from the hexagonal hub.

Further significant developments had to await the unraveling of atomic structure that was to take place at the end of the nineteenth and the beginning of the twentieth century. Scientists realized that the electrons surrounding the nucleus were grouped in shells, with all the electrons in a shell being the same distance away from the nucleus. An American chemist, Gilbert Lewis, proposed that the chemical activity of elements depended upon how close their outer shells were to being filled with electrons. Inert gases such as helium were chemically inactive because their outer shells were filled, while elements that had only one electron (or were missing one electron) in their outer shell were highly active. This led to the concept of the ionic bond, in which an element with an excess of electrons in its outer shell donates those electrons to an element with a deficiency of electrons in its outer shell. Another important type of chemical bond, the covalent bond, involves the mutual sharing of electrons between elements.

In the 1930s, Linus Pauling applied the new quantum mechanics to developing the theory of the chemical bond. Pauling showed that two elements could reduce their energy by forming a chemical bond, but that this would only take place if the atoms were close to each other. To reduce a chemical compound to its constituent elements, Pauling's mathematics showed that it was necessary to add energy, a fact known to the ancients who smelted iron from iron ore. Pauling's theory explained additional properties of chemical bonds and chemical reactions beyond the reach of previous theories. For this work, he was awarded the Nobel Prize for chemistry in 1954.

To gain a true appreciation of the respect in which Pauling was held by other scientists, read *The Double Helix* by James Watson, which describes the struggle to unravel the structure of DNA. Pauling had just deciphered some of the basic structure of proteins, and one of the highlights of the book is Watson's description of the relief he experienced when he realized Pauling's hypothetical structure for DNA contradicted recently taken X-ray diffraction photos of which Pauling was unaware. The impression one gets is that of a gunslinger in Dodge City who realizes that the fastest gun in the West is out of ammunition.

I was fortunate to hear a PBS broadcast of a Pauling question-and-answer session with some students. The first thing I noticed was that Pauling paused for about five seconds, and then answered every question in complete sentences. One question that he was asked by a student was what he thought of astrology. Pauling replied that the basics of astrology were devised by Ptolemy, who was undoubtedly one of the most brilliant men of his time, but was hampered by the lack of any sort of accurate data. Pauling said that he felt that Ptolemy, were he alive today, would be quick to reject astrology as completely unscientific.

However, my favorite Pauling answer did not take place at this session. He was once asked how he got so many good ideas, and replied that he just got a lot of ideas, and threw out the bad ones.

CATALYSIS AND PHYSICAL CHEMISTRY

One of the unwritten rules that many scientific graduate students learn is that it doesn't pay to be too brilliant when you are writing a thesis. There is a very good reason for this—in order to get to be a player in the science game, you first have to be accepted by the scientific community. Getting a doctorate requires doing something that the science community regards as good science, and many brilliant ideas require a lot of time before they are considered good science.

Svante Arrhenius had been a child prodigy. When the time came for him to write his thesis, he chose as his subject electrolytes, substances that were capable of conducting electricity when dissolved. Arrhenius proposed that when a molecule of an electrolyte actually dissolved, it separated into charged particles called ions, which enabled the current to flow. At the time, chemists adhered to Dalton's picture of the atom as indivisible, and so Arrhenius's theory was rejected by the majority of the scientific community. His thesis, however, was given the lowest possible passing grade, possibly on the ground that even though it was obviously erroneous, it was undoubtedly brilliant.

When a phenomenon is not completely understood, scientists may have a majority opinion, but there is usually a skeptical minority. Arrhenius sent copies of his thesis to several of the leading chemists, one of whom was Friedrich Ostwald. Ostwald was convinced of the validity of Arrhenius's ideas and helped to spread them, even as Arrhenius was continuing to gather evidence to support his views.

Gradually, the chemists became increasingly convinced of Arrhenius's ideas, but the key development in establishing them occurred when J. J.

Thomson identified the electron, a subatomic particle, and when Henri Becquerel showed that radioactivity involved the breakdown of atoms.

Many of the ideas of both Arrhenius and Ostwald lay at the junction of both physics and chemistry; indeed, the two scientists practically created the subject of physical chemistry. An excellent example of the importance of physical chemistry can be seen in the process of catalysis. Arrhenius realized that chemical reactions usually proceed more rapidly when the reacting substances are heated. We see this every day in the kitchen; it takes a shorter time to cook a roast at a high temperature than at a low one. Arrhenius suggested that molecules needed to be supplied a certain amount of energy, the "energy of activation," in order to participate in a chemical reaction.

Ostwald, meanwhile, was busily applying Arrhenius's ideas on ionization to a different aspect of catalysis. Some chemical reactions, such as the production of sugar from starch, are catalyzed by the presence of an acid. Ostwald realized that this type of catalysis involved lowering the energy of activation of the reacting substances.

Ostwald's observations on catalysis immediately found application in industry. Ostwald himself helped devise a procedure using platinum as a catalyst in making nitric acid more efficiently. Because nitric acid is important in the manufacture of high explosives, Ostwald's process enabled Germany to produce explosives during World War I without importing raw materials. The contribution of Ostwald to lengthening World War I must be counterbalanced by the fact that Ostwald's theories of catalysis were later to help explain the activity of enzymes, and so Ostwald indirectly contributed to the development of biochemistry and genetic engineering.

It is rather paradoxical that many of the contributions of Arrhenius and Ostwald could be explained by the atomic theory of Dalton, yet Ostwald himself did not believe in that theory until he was more than fifty years old! Perhaps this explains the fact that Ostwald was such a strong early supporter of Arrhenius's ideas on ionization, which required the breakdown of atoms if one believed in the atomic theory. Since Ostwald didn't believe in the atomic theory, this might have made it easier for him to agree with Arrhenius.

The Foundations of Biochemistry

It is chemistry that makes life possible.

We do not yet know how this was first achieved, but we do know that practically all the processes that characterize life are chemical reactions. Life must exhibit homeostasis—the ability to maintain a relatively stable internal environment. It must be able to convert food to energy. It must be able to reproduce, and do so relatively faithfully, with only minor changes (if any) from generation to generation. Each of these processes is a remarkable complex of chemical reactions. Deciphering these reactions and their implications constitute biochemistry, which is arguably the most important branch of science in promoting the continued well-being of humanity, both as individuals and as a species.

THE SYNTHESIS OF UREA

The theory of evolution has been under constant siege from the moment it was first propounded, and attempts have been made in many states either to have it removed entirely from the curriculum, or to denigrate its status. One strategy has been to require that creationism, an alternate view of the Universe in which everything was created by a supreme being, must be taught as well as evolution, placing the two theories on an equal footing.

In order to consider a suit by creationists to require the teaching of creationism, a judge undertook the reasonable task of trying to discover exactly what constitutes a scientific theory. He finally settled on a definition of scientific theory from the philosopher Karl Popper: a scientific theory is one that can be falsified. Experiments can be performed, or evidence can be found, which demonstrate that the theory is false. Under that definition, creationism is not a scientific theory because it is impossible to perform an experiment or find evidence that will invalidate the fundamental hypothesis of creationism.

Georg Stahl, a German who was a contemporary of Isaac Newton, was a pioneer in the fields of chemistry and biology. He observed, he experimented, he theorized, and he came up with two theories that were to have a profound effect on the development of chemistry. As we have seen, his phlogiston theory on the nature of combustion was shown to be false by Antoine Lavoisier in the latter portion of the eighteenth century. His other theory, vitalism, held on substantially longer.

Vitalism can be summarized by saying that there are two sets of laws: one governing inanimate objects and one governing living things. At the time it was propounded, vitalism represented an attempt to retain some of the mystic wonder with which religions viewed the phenomenon of life, while at the same time incorporating aspects of the newly emerging sciences. With the possible exception of perspiration, which consists primarily of the simple chemicals water and salt, nearly every chemical associated with life could not be analyzed by the techniques and instruments available in the eighteenth century. Just because you can't do something doesn't mean it can't be done, but it is very easy to accept that something can't be done because it hasn't yet been done. And until the beginning of the nineteenth century, the chemicals produced by living organisms resisted all attempt by scientists to analyze them.

As a result, vitalism remained a tenable theory as late as the early nineteenth century. The greatest chemist of the time, the Swede Jöns Jakob Berzelius, had in fact classified chemicals into two groups: inorganic and organic, organic chemicals being the products of life processes.

While debate raged in the chemical world as to whether inorganic and organic compounds were subject to different laws, the German chemist Friedrich Wohler was engaged in experiments with inorganic cyanide compounds. In 1828, he heated some ammonium cyanate. The residue of this experiment were crystals that seemed suspiciously familiar to Wohler. Subsequent testing revealed that the crystals were urea, which is the primary chemical compound present in urine. An organic compound had been created in the laboratory from inorganic precursors.

Chemistry today is still divided into organic chemistry and inorganic chemistry, although the distinction is now based on the chemistry of the element carbon rather than on the origins of the chemical. Despite the fact that the vitalistic theory had been overturned, it was still a key step on the road to scientific knowledge, which often emerges from initial ventures in precisely the wrong direction.

Without attempting to disparage the significance of Wohler's experiment, some historians have argued that the ammonium cyanate he used was actually an organic compound, and so his experiment did not really transform inorganic substances into organic ones. Even conceding this point, Wohler's result convinced chemists that the distinction between organic and inorganic chemistry was an artificial one, and paved the way for the explosion in synthetic organic chemistry that was to come. Incidentally, Stahl might be pleased to note that vitalism lives on—although there is no chemical difference between synthetically produced vitamins and those obtained from natural sources, many people are still willing to pay a premium for the latter.

THE ISOLATION OF ZYMASE

In vino, veritas—in wine, there is truth. The discovery of the truth of the almost-miraculous procedure that turns grape juice into wine is a story that reaches far back into history, and climaxes with the birth of biochemistry in the nineteenth century.

Wine has been around for almost ten thousand years, and its manufacture amazed the ancients. In fact, so astounding was the transformation of grape juice to wine that, in the Middle Ages, it effected a transformation on the chemical theory of the time, which held that the world was comprised of four elements: earth, air, fire, and water. The process by which grape juice became wine must have involved a *quinta essencia*, a fifth element, which reflected and shaped the unique form of living matter. This fifth element uniquely characterized the life that possessed it, and our word "quintessential" reflects this characterization.

Fermentation, the process by which grape juice becomes wine, continued to be studied by many of the great scientists of the eighteenth and nineteenth centuries. The brilliant French chemist Lavoisier showed that the addition of a small amount of yeast to a sugar solution resulted in the production of alcohol, thus classifying fermentation as a chemical reaction. Since this reaction did not occur without the yeast, it was clear that the yeast played a vital role in the process.

But what kind of a role was yeast playing? The German chemist Justus von Liebig, one of the founders of organic chemistry, held that the role of the yeast was to emit vibrations that accelerated the chemical reactions. At approximately the same time, the German biologist Theodor Schwann and the French inventor Charles Caignard de la Tour discovered that yeast was actually a living entity. This led to the biological theory of fermentation, in which sugar was ingested by the yeast, and alcohol and carbon dioxide excreted.

Louis Pasteur, one of the undoubted titans of science, had helped rescue a tottering French wine industry by his discovery that bacteria could spoil wine. Pasteur's investigations resulted in the realization that yeast was composed of living cells, and that the transformation of sugar into alcohol and carbon dioxide was an integral part of their life. However, Pasteur believed that this transformation depended upon the living condition of the yeast. This was a somewhat unusual conclusion to reach for the man who had effectively destroyed the vitalistic theory of spontaneous generation—that life could simply arise from nonliving matter.

The ultimate resolution of the question came in 1897. Two brothers, Hans and Eduard Buchner, through a series of carefully designed and constructed experiments, managed to isolate zymase, the enzyme responsible for transforming sugar into alcohol. Crushing the yeast cells through filter paper, they obtained an extract that, although clearly not alive, was able to convert sugar to alcohol. This experiment led eventually to the realization that the processes of life were essentially chemical in nature, and that the role of enzymes, of which zymase was one, was to accelerate the chemical transformations that enabled living cells to transform raw materials into usable products.

Different sciences arise in different ways. Some are the result of a single dazzling insight, such as Mendel's formulation of the concept of the gene. Others are the result of long years of observations and theorizing, false trails and dead ends. Once the correct path is determined, however, the advances often come with dazzling swiftness.

Although zymase was not the first enzyme to be discovered, it was the first whose actions were observed in vitro (literally, "in glass"), without the necessity of the participation of a living entity. A few years after the isolation of zymase, Franz Hofmeister formulated the central dogma of biochemistry, that all cellular processes would be shown to be controlled by enzymes.

Hans Buchner died in 1902, and was thus unable to share in the Nobel Prize awarded to Eduard in 1907. Eduard continued a distinguished career as a professor of chemistry until, in 1917, at the age of fifty-seven he volunteered for a second tour of duty in World War I. Captain Eduard Buchner was wounded on the Eastern Front, and died two days later.

THE STRUCTURE OF INSULIN

It was apparent to the chemists of the late eighteenth century that the properties of a substance depended upon its molecular composition. What was not immediately apparent was that the properties of a substance also depended upon the molecular architecture.

If one thinks of a house, for example, it is obvious that a house built of brick will have different properties from a house built of wood. However, two houses may be built from precisely the same number of bricks and be radically different—the architecture can create very dissimilar buildings. One may be light and airy, the other somber and foreboding.

Hints that the same idea might permeate chemistry began to crop up early in the nineteenth century. One of the first to get an insight into this was Louis Pasteur.

Pasteur's first accomplishment as a scientist was to show that crystals of tartaric acid consisted of two distinct types. One type of crystal polarized light so that it bent to the right, and the other polarized light so that it bent to the left. News of this discovery reached Jean-Baptiste Biot, one of France's greatest scientists. Biot had spent considerable time investigating the polarization of light, a phenomenon that concerns a type of organized behavior displayed by light waves, and is used to create reduced-glare sunglasses, as well as to detect lines of stress in materials. (Lasers involve a different type of organized behavior of light waves.) Biot was skeptical, and asked Pasteur to demonstrate this for him. Pasteur prepared the experiment, and when Biot observed the desired effect, he grabbed Pasteur's hand and said, "My dear fellow, I have been so enamored of science all my life that this causes my heart to beat faster."

By the end of the nineteenth century, chemists were certain that proteins, the most crucial of the organic molecules, were constructed by linking together fundamental structural units known as amino acids. It was later discovered that the genetic code in DNA gives instructions as to which amino acids, and in which order, are to be used in the making of proteins. With the advent in the 1940s of a technique called chromatography, it became possible to work out the amino acid composition of a protein. However, one cannot understand protein function by knowing its amino acid composition any more than one can tell what type of house is being built simply by knowing how many of which type of brick are used. The architecture, in proteins as in houses, is all-important.

Frederick Sanger, a British biochemist, was the first person to successfully unravel the structure of a protein. He worked with the protein insulin, which was known to consist of fifty amino acids. The fifty amino acids were assembled in two chains, one nineteen amino acids long, the other thirty-one. Sanger had discovered a chemical that would attach itself to a specific end of a chain of amino acids. Using this chemical (now known as Sanger's reagent) and the technique of chromatography, Sanger would break the big chains into fragments, painstakingly work out the location and structure of each fragment, and then fit the fragments together, much as one puts together a jigsaw puzzle by assembling individual pieces into small portions, and then fitting the portions together. It was arduous work, extending over eight years. However, by 1953 he had worked out the structure of the insulin molecule, an epochal achievement for which he was awarded the 1958 Nobel Prize in chemistry.

The importance of structure, as opposed to mere composition, cannot be underestimated. For example, the reason that carbon monoxide poisoning occurs is that the carbon monoxide molecule fits into an indentation of the hemoglobin molecule even more snugly than oxygen. As a result, faced with a choice of grabbing oxygen or carbon monoxide, the hemoglobin molecule will go for the carbon monoxide, with fatal results. So strongly does hemoglobin sequester carbon monoxide that simply breathing oxygen is not an adequate treatment for carbon monoxide poisoning—it is necessary to breathe oxygen at two to three times atmospheric pressure to force the hemoglobin to disgorge the carbon monoxide molecules.

As evidence of the importance the scientific community attached to Sanger's achievement, it is worth noting that 1953, the year in which Sanger deciphered the structure of insulin, was also the year that Watson and Crick deciphered the structure of DNA. Watson and Crick, however, did not receive their Nobel Prizes until 1962.

CHAPTER 4

Matter

Complicated stuff is made of simple stuff, and one of the great quests of science has been to understand what the simple stuff is, and how it works. Chemistry is concerned with how the simple stuff fits together to form complicated stuff, but physics is concerned with what the simple stuff is.

Phases of Matter

There are three fundamental phases of matter—solid, liquid, and gas. Although the Greeks didn't explicitly state it in this fashion, they believed the world was constructed of four basic elements—earth, air, fire, and water. The three fundamental phases are represented here—earth is solid, water is liquid, and air is gaseous. It would be millennia before the atomic theory revealed precisely what made water a solid (as ice), a liquid (as water), and a gas (as steam).

A fourth phase of matter—plasma—was discovered in the late nineteenth century, but one could make a pretty good argument that the Greeks had at least a hint of this, because fire—which generally results from heat energy being applied to a combustible material—is in some sense analogous to plasma, which consists of ionized particles and generates its own magnetic field. Fire, too, consists of particles generating a form of energy.

Water, being the most common substance available—at least, available in all three phases of matter—was a natural candidate for investigation. One could see that as winter came, water in rivers and lakes transformed into ice, and back to water again with the coming of spring. Similarly, heating water causes it to become steam, which became water again when it condensed on a cooling surface. But these were merely qualitative descriptions of phenomena. While science often starts with qualitative descriptions, it really only acquires significant predictive power—and the utility associated with this predictive power—when these qualitative descriptions become quantitative. And in order for them to become quantitative, there have to be instruments available capable of measuring parameters. The Greeks could measure volume and mass, but had no way to measure parameters such as temperature.

THE GAS LAWS

Hero of Alexandria was a Greek mathematician, scientist, and engineer who actually made it into the history books under two different names. He was also known as Heron, and Heron's formula in mathematics enables one to compute the area of a triangle from the length of its sides. But he is most famous as an engineer and scientist—he constructed both windmills and steam engines, although nowadays these would probably be classified as proof-of-concept prototypes rather than as industrial equipment ready for prime time. Hero also made an intensive study of air. He showed that air was actually a substance (unlike fire) by demonstrating that water would not enter a container filled with air unless the air was first removed. Hero also argued that, because air was compressible, it must consist of particles separated by empty space—a remarkable insight.

Almost 1,800 years would pass before as talented an engineer would again investigate air. The engineer was Otto von Guericke, who invented the first air pump. It cost him a prodigious amount of money, but von Guericke was also a good promoter, and gave impressive demonstrations on the nature of a vacuum. He showed that candles would not burn and animals could not live in a vacuum, and demonstrated the power of air pressure by evacuating a metal sphere and showing that teams of horses could not pull it apart.

The British scientist Robert Boyle read of von Guericke's experiments, and decided to duplicate them. He actually improved von Guericke's air pump, and used it to demonstrate the truth of Galileo's contention that, in a vacuum, a feather and a lump of lead should fall at the same rate (this experiment was also demonstrated somewhat more dramatically some three hundred years later—on the surface of the Moon). Boyle also showed that sound could not be heard in a vacuum, but that an electric charge could jump across one.

As a result of these experiments, Boyle was led to investigate the nature of gases. He was the first chemist to actually collect a gas, but his major contribution was to measure the relation between the pressure applied to a gas and the volume that it occupied. Using a J-shaped tube, Boyle demonstrated that the volume of the gas was inversely related to the pressure. Double the pressure and the volume halved; triple the pressure and the volume was reduced to a third of the original volume. This relation is known as Boyle's law.

A century later, the French balloonist Jacques Charles was also interested in gases, although from a more practical point of view—he was, after all, a balloonist, and balloons are filled with hot air. Charles was the first balloonist to ascend to an altitude of more than 3,000 meters (about 10,000 feet). He was able to accomplish this feat because he had read of Cavendish's discovery of the much lighter hydrogen, and he realized that the lifting power of hydrogen would be far greater than that of air.

Charles's chief contribution was to show the effect of temperature on the volume of a gas. Again, this is perhaps not surprising in view of the balloonist's interest in heated gases. He discovered that different gases all expanded the same fraction of their initial volume when the temperature was raised by a given amount. For each degree Celsius that the temperature rose above 0°C, the volume increased by 1/273; for each degree the temperature fell below 0°C, the volume decreased by 1/273. In retrospect, this can be seen as foreshadowing the concept of absolute zero—if you keep reducing the volume by 1/273 for every degree the temperature is lowered, lower it 273 degrees and the volume is zero, so you can't lower it any more. And indeed, absolute zero is almost exactly 273 degrees below 0°C.

Charles did not actually publish this result, but it was later discovered (and published) by Joseph Gay-Lussac, who was by a curious coincidence a fellow balloonist! Ballooning was not only a passion with Charles, and the source of his scientific reputation, it also saved his life. He had the bad luck to be in the Tuileries when it was invaded by an angry mob during the French Revolution, but he had the presence of mind to recount some of his ballooning anecdotes to the bloodthirsty mob that accosted him. He must have been an exceptional storyteller, or had some truly fascinating anecdotes, as they let him go.

The Realm of the Atom

As Richard Feynman observed, the atomic theory was the key to many of the major scientific developments. By the middle of the nineteenth century, it was widely accepted that there were a number of basic elements such as carbon, iron, and oxygen, and that all the stuff on Earth was made up of these. But what about the heavens?

Meteors had been observed as far back as prehistory, but they were felt to be connected with the atmosphere—the word "meteor" is derived from the Greek word for "atmospheric." It wasn't until the great meteor shower of November 1833 that it was realized that the origin of meteors was extraterrestrial. Prior to that, in 1807 Professor Benjamin Silliman of Yale University had performed a chemical analysis of a meteor and shown that it contained iron, so it was known that at least some of what made up the heavens was made of the same stuff that made up the Earth. But the big breakthrough occurred half a century later.

SPECTROSCOPY

In 1835, the philosopher Auguste Comte attempted to go where no philosopher had gone before. Hitherto philosophers had tried to define the limits of human knowledge, but had generally done so by looking at moral, ethical, and religious questions that they felt would never be resolved. Comte compiled a list of questions he felt science would never be able to answer. One of those questions was to determine the composition of the stars. This was not an unreasonable idea; in 1838 Friedrich Bessel would show that the distance to the star 61 Cygni was six light-years, a distance far larger than anyone had even suspected.

This prediction probably went unnoticed by Robert Bunsen, a German chemist absorbed in studying organic arsenic-containing compounds. Bunsen was, perhaps, seventy-five years too early, for he was studying compounds that would not attract much interest until Paul Ehrlich developed a chemotherapeutic cure for syphilis from one such compound. These compounds were highly toxic, and Bunsen lost an eye and twice almost died from arsenic poisoning. Upon recovering, he decided that discretion was the better part of valor, and never again touched organic chemistry.

He then switched to a study of the role of heat in chemical reactions, inventing numerous devices. Interestingly enough, he did not invent the one for which he is best known, the Bunsen burner, but he made it a standard tool in chemical laboratories.

One of his early students was Gustav Kirchhoff. The two worked together for four years, and then their paths diverged. Kirchhoff was initially interested in electricity (he was the first person to demonstrate that electricity moved at the speed of light), but when he and Bunsen were reunited, Kirchhoff took up Bunsen's investigation of photochemistry (chemical reactions that absorb or emit light). Bunsen at the time was using colored filters, but Kirchhoff, who had an extensive mathematical background and was an admirer of Newton, suggested that they use a prism, as had Newton during his investigation of optics.

The two combined Thomas Young's idea of passing light through a slit with Newton's idea of passing light through a prism. Thus was born the spectroscope. The Bunsen burner was used to heat chemicals to incandescence, and the light passed through a spectroscope to throw a pattern of colored lines on a screen. It was soon discovered that this pattern of colored lines was a chemical fingerprint, and that each element had its own characteristic pattern, or spectrum.

Using the spectroscope to analyze the light of the sun, Kirchhoff discovered a spectral line characteristic of the element sodium. Since there was no sodium in the Earth's atmosphere, and certainly none in the vacuum between the sun and the Earth, the conclusion was inescapable: sodium existed in the sun. The same technique would later be used on the light from stars, enabling their chemical composition to be determined.

Kirchhoff's banker, a pragmatic sort, asked Kirchhoff, "Of what use is gold in the sun if I cannot bring it down to Earth?" Kirchhoff was soon to be awarded a medal and an accompanying monetary prize from Great Britain for his work. The money was awarded in golden sovereigns. Kirchhoff, handing them to his banker, could not resist the opportunity to remark, "Here is gold from the sun."

Spectroscopy has advanced considerably since its initial discovery. There are instruments on the James Webb telescope, launched in December 2021, capable of detecting the signatures of molecules in the atmospheres of exoplanets. At the outer limits of which molecules can be detected by the instruments on the Webb telescope are oxygen and ozone, two molecules that on Earth indicate the presence of life. The SETI project (Search for Extraterrestrial Intelligence) has scanned the Universe for more than half a century, trying to find a signal that might indicate the presence of intelligent life. Maybe SETI is asking for too much, but if the Webb telescope finds oxygen or ozone in the atmosphere of an exoplanet, it would be a likely indicator that life exists elsewhere in the Universe.

Auguste Comte, the author of the ill-fated prediction about the stars, later went insane, and died two years before the development of the spectroscope. Comte's prediction has been treated by some historians as the work of a buffoon, but the truth is that Comte was right in principle, if wrong in specifics. Almost a century after Comte's prediction, the German physicist

Werner Heisenberg would formulate the uncertainty principle of quantum mechanics, a fundamental discovery showing that Comte was right, and that there are limits to scientific knowledge. A few years later, Kurt Gödel would prove his famous incompleteness theorem, showing that there were limits to mathematical knowledge as well.

THE STRUCTURE OF THE ATOM

With the atomic theory firmly in place by the end of the nineteenth century, it was thought that the ultimate constituents of matter had been discovered. The prevailing view of what an atom would prove to be, if indeed it were ever possible to actually see one, was that it would be a small, hard, featureless sphere. Meanwhile, the nature of electromagnetic energy occupied the attention of several physicists.

James Maxwell had shown that electromagnetism could be regarded as waves, but in the late 1870s, William Crookes discovered that the rays emitted by the cathode of a vacuum tube could be deflected by a magnetic field. This convinced Crookes that these rays were actually particles carrying an electric charge. Two decades later, J. J. Thomson was able to show that these rays could also be affected by an electric field, and were therefore definitely particles. By careful experimentation, Thomson was able to go even further, measuring the charge-to-mass ratio of the particles. From this he deduced that the particles, soon to be known as electrons, were extremely small, having approximately 1/1837 the mass of a hydrogen atom. Since the hydrogen atom was the smallest possible atom, electrons were substantially smaller than atoms. The field of subatomic physics had dawned.

One of Thomson's assistants was Ernest Rutherford, who had lived the early portion of his life on a potato farm in New Zealand. Rutherford received news of the offer of a scholarship at Cambridge while digging potatoes on his father's farm. Flinging aside his shovel, he declared, "That's the last potato I'll dig," and set sail for England.

Rutherford was initially interested (as was practically everyone else) in the emissions given off by radioactive material. He embarked upon the study of how these rays bounced off thin sheets of metal. He fired alpha particles, which had a positive electric charge, at a sheet of gold only two thousand atoms thick. When some of the alpha particles bounced almost straight back, Rutherford realized that there had to be a concentrated region of positive charge somewhere in the atom, as it would take a positive charge to repel the positively charged alpha particles. However, since the vast majority of the alpha particles passed straight through the sheet of gold without being deflected at all, Rutherford realized the atoms must be mostly empty space. He therefore proposed a model of the atom in which the positive charges (protons) in the nucleus were packed in tightly around the center, surrounded by electrons in the outer layers.

One of Rutherford's assistants was Niels Bohr, a young Danish scientist who had migrated to Rutherford after working with Thomson. The atomic model proposed by Rutherford had a number of deficiencies, one of which was its inability to explain why each atom had its telltale fingerprint of spectral lines. Using the newly developed quantum theory, Bohr made a radical assumption—that the electrons circled the nucleus like planets circling the sun. Moreover, quantum theory required that the orbits of the electron could only occur at certain specific distances from the nucleus—"in-between" orbits simply were not allowed. Bohr's theory explained the spectral lines occurring in the hydrogen atom, and Bohr's picture of the atom is, with some modifications, basically the one that is held today.

In the early twentieth century, the best way to assure yourself of being on the short list for a Nobel Prize was to be one of Thomson's assistants. Thomson himself received a Nobel Prize, as did Rutherford, Bohr, and five other Thomson assistants. Rutherford and Bohr also did yeoman service in helping Jewish scientists escape from Nazi Germany. When Denmark fell to the Germans, Bohr helped most of the Danish Jews escape Hitler's death camps. In 1943, Bohr himself fled Denmark to Sweden, then flew in a tiny plane to England, nearly dying en route from lack of oxygen. From there he went to the United States, where he was one of the physicists at Los Alamos who helped to develop the atom bomb.

THE QUANTUM HYPOTHESIS

As the nineteenth century came to a close, physicists around the world were beginning to feel their time had come and gone. The eminent physicist Philipp von Jolly advised his students to pursue other careers, feeling that the future of physics would consist of the rather mundane task of measuring the physical constants of the Universe (such as the speed of light) to ever-increasing levels of accuracy.

Still, there were minor problems still unresolved. One of the unsettled questions concerned how an object radiates. When iron is heated on a forge, it first glows a dull red, then a brighter red, and then white; in other words, the color changes in a consistent way with increasing temperature. Classical physics was having a hard time accounting for this. In fact, the prevailing Rayleigh–Jeans theory predicted that an idealized object called a blackbody would emit infinite energy as the wavelength falling on it became shorter and shorter. Short-wavelength light is ultraviolet; the failure of the Rayleigh–Jeans theory to predict finite energy for a radiating blackbody exposed to ultraviolet light came to be known as the "ultraviolet catastrophe."

The Rayleigh–Jeans theory operated under a very commonsense premise—that energy could be radiated at all frequencies. An analogy would be to consider the speed of a car; the car should be able to travel at all velocities up to its theoretical limit. If a car cannot go faster than 100 miles per hour, for instance, it should be able to move at 30 miles per hour, or 40 miles per hour, or 56.4281 miles per hour.

One day in 1900 the German physicist Max Planck, one of those advised by von Jolly to consider another major when he entered the university, made a bizarre assumption in an attempt to escape the ultraviolet catastrophe. Instead of assuming that energy could be radiated at all frequencies, he assumed that only certain frequencies were possible. Continuing the analogy with the speed of the car, Planck's hypothesis would be something like only speeds that were multiples of 5, like 25 miles per hour, 40 miles per hour, and so on would be possible. He was able to show almost immediately that this counterintuitive hypothesis resolved the dilemma, and the radiation curves he obtained matched the ones recorded by experiment. That day, while walking with his young son after lunch, he said, "I have had a conception today as revolutionary and as great as the kind of thought that Newton had."

His colleagues did not immediately think so. Planck was a respected physicist, but the idea of the quantum—energy existing only at certain levels—was at first not taken seriously. It was viewed as a kind of mathematical trickery that resolved the ultraviolet catastrophe, but did so by using rules that the real world did not obey. Planck's idea languished for five years, until Einstein used it in 1905 to explain the photoelectric effect. Eight years later, Niels Bohr used it to explain the spectrum of the hydrogen atom. Within another twenty years, Planck had won a Nobel Prize, and quantum mechanics had become one of the most fundamental theories of the real world, explaining the behavior of the world of the atom and making possible many of the high-tech industries of today.

With the coming of the Nazis, German science suffered severely. Many of the leading scientists were either Jewish or had Jewish relatives, and fled the country. Many others reacted with abhorrence to the Nazi regime, and also departed. Planck, although deploring the Nazis, decided to stay in Germany. It was to be a tragic decision. In 1945, Planck's younger son was executed for his part in the "Revolt of the Colonels," the unsuccessful attempt by several members of the German armed forces to assassinate Hitler.

RADIOACTIVE DECAY AND ISOTOPES

When John Dalton first proposed his atomic theory, the atoms he envisioned were immutable, and the atoms of any particular element were identical in shape, size, and mass. As the twentieth century began to unfold, and scientists became able to discern the structure of the atom, the validity of these hypotheses became open to doubt.

The first of Dalton's atomic characteristics to tumble was the idea that atoms were immutable. Interestingly enough, the hypothesis of immutability was one of the most noteworthy differences between Dalton's atomic theory and the ideas of the old alchemists. The alchemists felt that the elements could be transformed into one another, and a great deal of effort was expended in a futile search for the "Philosopher's Stone," whose touch would change lead into gold. In 1902, Ernest Rutherford and Frederick Soddy conducted a series of experiments with the newly discovered radioactive materials. They showed that radioactive elements were subject to spontaneous decay; radioactivity consisted of an emission of particles whose absence actually transformed the radioactive element into a different element. With remarkable insight, Rutherford and Soddy suggested that this transformation took place at a subatomic level. Their discovery necessitated a revision in the atomic theory, and was eventually to lead to Rutherford's views on the existence of an atomic nucleus.

Soddy spent the most productive portion of his career working on phenomena associated with radioactive decay. He helped discover that lead was the end product of all radioactive decay series. He also devised a law to explain radioactive decay, called the radioactive displacement law.

At the time, scientists had noted that two different types of particles could be emitted during radioactive decay, which they had named alpha and beta particles. Soddy observed that when an alpha particle was emitted, both the atomic weight of the element from which the particle was emitted and the nuclear charge decreased by two. With the aid of hindsight, we can see that this is explained if an alpha particle consists of two protons and two neutrons, but the neutron would not be discovered for almost twenty years. When a beta particle was emitted, the atomic weight did not change, but the nuclear charge increased by one. It would later be understood that this occurred because when a neutron decayed, it formed a proton and ejected an electron.

Soddy realized that the same element could be formed from an element two places higher in the periodic table by alpha decay, or by an element one place lower in the periodic table by beta decay. However, the atomic weights as measured didn't add up. Soddy interpreted this as showing that elements could exist in more than one form, which he called isotopes. Two different isotopes of the same element would have the same nuclear charge, but different atomic weights.

This was another revision to Dalton's atomic theory. Atoms, no longer immutable, were now no longer identical. The discovery of the neutron by British physicist James Chadwick eventually supplied the mechanism by which different isotopes were able to exist, as neutrons did not alter nuclear charge, but did change the atomic weight.

Two different isotopes of the same element have essentially the same chemical properties but different atomic properties. Different isotopes have been used in medicine, and in radioactive dating. The most historically significant use of different isotopes was discovered in the late 1930s. Ninety-nine percent of naturally occurring uranium has an atomic weight of 238, and less than 1 percent has an atomic weight of 235. When bombarded with neutrons, the less stable uranium-235 fissions into two smaller atoms, releasing both energy and additional neutrons. These additional neutrons then slam into other uranium-235 atoms, which in turn release more energy and more neutrons. This process is called a chain reaction and is the mechanism behind the explosive power of the atomic bomb.

X-RAY CRYSTALLOGRAPHY

New Age adherents are only the latest of a long line of people to be fascinated by crystals. The beauty and symmetry of many forms of crystals, to say nothing of their rarity, have caused them to be highly valued for thousands of years, as well as being suspected of having mystical properties.

The scientific investigation of crystals can be traced to Nicolaus Steno, a seventeenth-century Danish scientist who was one of the first to suggest that fossils were the petrified remains of long-dead animals. Steno observed that when a crystal broke, it did so not in random pieces, but in straight planes that always met at characteristic angles. This observation would later become known as the first law of crystallography.

Crystallography languished for a century until René Haüy, a French priest, accidentally dropped a piece of calcite that was part of a friend's mineral collection. Apologizing for his clumsiness, he picked up the pieces and noted, as had Steno a century earlier, that the fragment faces showed straight planes meeting at a constant angle. Haüy pursued this observation further, this time deliberately breaking crystals. He suggested that crystals were composed of "unit cells," formed from repeating geometric configurations—a surprisingly prescient prediction of the atomic structure that crystals display.

The nineteenth century witnessed extensive investigation into other properties of crystals. Certain crystals were discovered to have the ability to polarize light. Another unusual property, discovered by Pierre Curie, is piezoelectricity—squeeze certain types of crystals, and they emit an electric current. Conversely, an electric current will cause certain crystals to vibrate. This principle is the basis of the quartz watches that were popular half a century ago. In the latter portion of the twentieth century, the LCD—liquid crystal display—has made possible inexpensive timepieces of unprecedented accuracy.

Perhaps the most important scientific result from the investigation of crystals occurred because of a controversy surrounding the nature of X-rays. Were they particles, like cathode rays, or longitudinal waves, like sound, or transverse waves, like light? By 1910, the balance of opinion had shifted to the third point of view, but the problem lay in measuring the wavelength of X-rays.

With ordinary light, the wavelength was measured by a diffraction grating, in which light was bounced off a piece of metal having lines that were separated by precise distances etched in the metal. The shorter the wavelength of the light, the closer the lines had to be, and it simply was not possible to make the lines close enough to diffract the X-rays.

Max von Laue, a German physicist who had been Max Planck's assistant, realized that, in crystals, nature had already manufactured the perfect diffraction grating for X-rays. In 1912, he shone X-rays on a crystal of zinc sulfide, and the diffracted X-rays were bounced onto a photographic plate, where they could be detected. The regular crystal pattern of the zinc sulfide was revealed by the spots of light where the X-rays had struck the plate.

So X-ray crystallography could be used to determine the wavelength of an X-ray if the spacing of a crystal were known. Far more important, though, was that the reverse was also true. If the wavelength of an X-ray is known, it can be used to determine the structure of a crystal. This technique has been one of the most important tools in discovering the structure of complex substances.

Two extremely significant developments in unearthing the structure of complicated substances have been greatly aided by two of the best X-ray crystallographers—Rosalind Franklin and Dorothy Crowfoot Hodgkin. Franklin's X-ray studies of crystals of DNA molecules were critical in helping Watson and Crick determine the structure of DNA. During World War II, the incredible antibiotic properties of penicillin were realized by the British government, but the difficulty in procuring it from natural sources made it the most expensive substance on Earth. Hodgkin used X-ray crystallography and a primitive electronic computer to discover the structure of penicillin. In so doing, she made possible its synthesis, and the resulting mass production has saved millions of lives.

ATOMIC NUMBERS

Long before the concept of a "Dream Team" had come into existence, the great experimental physicist Ernest Rutherford had assembled his version of one. Rutherford, his coworkers, and students represented the cream of the early-twentieth-century crop of physicists, winning no fewer than seven Nobel Prizes.

None of those Nobel Prizes went to Henry Moseley, whom many feel might have been the greatest member of the Dream Team. Moseley joined Rutherford at Manchester University in 1910, after having excelled as a student at Eton and Oxford. After getting his feet wet studying beta emission from radium, Moseley then began an X-ray study of the elements.

X-rays had been discovered less than twenty years earlier by Wilhelm Roentgen, but already scientists were taking advantage of their extremely short wavelength to probe the structure of matter. Two pioneers in this area were the father-and-son team of William Henry Bragg and William Lawrence Bragg. The Braggs had recently discovered that when elements were excited by X-rays, the spectrum contained several bright lines that served as a signature of the element.

Any work done with elements was performed against the background of Mendeleev's periodic table of the elements. However, there were still problems with the exact structure of the periodic table. When Mendeleev originally compiled the table, he arranged the elements in increasing order of atomic weight. Although there was no way for Mendeleev to know it, arranging elements in this order was to introduce both errors and complexities into his periodic table. Because the atomic weight of an element depended on the isotopic composition of the element that Mendeleev had studied,

Mendeleev's order had occasionally been shown to be incorrect. Also presenting a problem was that no one could be sure two elements with a large gap in atomic weights were adjacent in the table, as a new element might be discovered that belonged between them.

Moseley decided to compare the atomic weights of the elements with the spectral lines that the Braggs had discovered. He discovered that there was a simple correspondence between the spectra and the atomic weights, as the wavelengths of the signature lines in the spectra increased in a regular pattern as the atomic weights increased.

Other physicists had suggested that there might be a correspondence between the number of protons in the nucleus and the atomic weight. Moseley solidified this idea by proposing that the number of protons in the nucleus actually accounted for the regular progression of the signature lines. He introduced the term "atomic number" to describe the number of protons in the nucleus of an element, and redefined the periodic table by arranging the elements in increasing order of atomic number.

This turned out to be the secret to determining the correct structure of the periodic table. It explained the elements that existed, and also predicted which elements would be discovered to fill existing holes. Finally, because atomic numbers are whole numbers, Moseley's theory predicted what elements would *not* be discovered, because elements cannot have fractional atomic numbers. The atomic number of hydrogen is 1 and that of helium is 2, so no element can be discovered between them.

Moseley reported his results in December 1913. In 1914 World War I began, and the patriotic Moseley enlisted in the Royal Engineers. He sailed for Turkey in June 1915, and lost his life to a sniper's bullet at the Battle of Gallipoli, a battle whose incredible loss of life was the result of stupid bungling. Moseley would undoubtedly have won a Nobel Prize for his work on atomic numbers, but Nobel Prizes are not awarded posthumously. The value of Moseley's work can be seen from the fact that Karl Siegbahn, a Swedish physicist who built upon Moseley's ideas, did win the Nobel Prize.

In researching this book, two individuals in particular left a profound impression on me. Henry Moseley died a senseless and unnecessary death at age twenty-seven as cannon fodder at the Battle of Gallipoli, and Rosalind Franklin, mentioned in the previous section, succumbed to cancer at age thirty-eight. No A. E. Housman has written an ode to a scientist dying young, but it seems to me that someone should, as we are all poorer for the discoveries they might have made.

ANTIMATTER

The 1920s marked an incredibly productive decade in physics. The revolution that was quantum mechanics had been under way for more than twenty years, achieving a significant triumph with Bohr's quantum-mechanical description of the hydrogen atom. In addition, 1919 had brought the confirmation of Einstein's theory of general relativity. All this activity attracted many bright young scientists to physics.

Louis de Broglie was a member of the French nobility. Nobility in France was not always an asset; de Broglie's great-great-grandfather had died by guillotine during the French Revolution. After receiving a degree in history, de Broglie worked in radio communication during World War I, and his interest turned to science.

Both Einstein and Planck had obtained important equations involving energy. Einstein's equation related energy to mass; Planck's related energy to wavelength. De Broglie did something almost childishly simple—he equated the two expressions, obtaining a relationship between wavelength and mass. Quantum mechanics had resolved the dilemma about light by showing that light was both wave and particle; could the same thing actually apply to physical objects?

The wavelength of an ordinary object, such as a grain of rice, was inconceivably short and could not be detected, but the wavelength of the smallest object then known, the electron, was long enough to be measured. In 1927 both Clinton Davisson and George Thomson managed to detect it. The wave-particle duality that applied to photons, which existed as energy, had now been shown to apply to electrons, which existed as mass.

Paul Dirac had begun his career as an electrical engineer, but found it difficult to get a job, and switched to mathematics. This was clearly the right move, as his talent was so evident that by the time he was thirty years old he had attained what was arguably the most prestigious post in the mathematical world, the Lucasian Professorship at Cambridge that had once been held by Newton. De Broglie's work had interested many mathematicians and physicists, and Dirac switched again, to mathematical physics.

This was yet another right move. Dirac derived equations that indicated that the electron could have two different types of energy states, one positive and one negative. If energy states were interpreted as electrical charge, the obvious interpretation of the negative energy state was the electron, which had a negative charge. Dirac's equations suggested that there must exist a particle identical in all ways to the electron, except that it would have positive charge.

Such a particle had never been detected, and Dirac's results were greeted with skepticism. However, events shortly vindicated Dirac. Carl Anderson was an American physicist who was studying cosmic rays, immensely energetic particles generated in interstellar space. These particles were so energetic that they could not be studied with ordinary cloud chambers. Anderson inserted a lead plate to slow down the particles, thus making them accessible to study. One day Anderson was examining the track of a particle that appeared to be identical to an electron, but where an electron curved in one direction in response to a magnetic field, Anderson's particle curved in the opposite direction. Anderson's particle was the first antimatter to be discovered, and it was indeed the one Dirac had discovered in his equations.

When matter and antimatter meet, they annihilate each other in a burst of energy. One of the prevailing mysteries of physics is why the Universe is mostly made of ordinary matter. Why don't entire galaxies made of antimatter, with antistars, antiplanets, and perhaps antipeople exist? In 1995 scientists created antihydrogen, an atom with an antielectron orbiting around an antiproton. The atom lasted for only a few millionths of a second before it was annihilated in a collision with ordinary matter.

NUCLEAR FISSION

Leo Szilard, who had been born in Hungary of Jewish parents, was one of the first to see the writing on the wall. A brilliant physicist whose work had taken him to a position on the faculty of the University of Berlin, he recognized that there was no future for him in Germany after Hitler came to power, and went to England. In 1934, while walking the streets of London, he invented the concept of the chain reaction. His original idea involved the fission of the metal beryllium into helium atoms, and although he could not demonstrate the process, he could describe it. Which he did, in the process taking out a patent on the procedure. As he could see the military potential inherent in the idea, he kept the patent secret.

Back in Germany, Lise Meitner felt relatively safe even though she was Jewish, as she was an Austrian national. She stayed in Germany in order to continue her research collaboration with Otto Hahn, with whom she was to work for more than thirty years. One of the earliest women to pursue a career in science, she had been the victim of antifeminist prejudices, as the director of the laboratory in which she worked initially refused to let her in the laboratory when men were working. She and Hahn were deeply involved in the process of studying the behavior of uranium under neutron bombardment when Hitler annexed Austria. Despite the fact that Meitner had served as a nurse in the Austrian Army in World War I, she knew that Germany was now extremely dangerous for her. Dutch scientists enabled her to come to the Netherlands without a visa, and Niels Bohr then helped her obtain a position in Sweden.

Fritz Strassmann replaced Meitner as Hahn's partner, and in early 1939 they published a paper describing the results of their experiments with uranium. Initially, they suspected that the bombardment had created the radioactive element radium, which is chemically similar to barium. As a result, they treated the bombarded uranium with barium. However, they could find no evidence of radium. When they published the results of their work, they carefully avoided the suggestion that the uranium atom had fissioned, with the lighter barium as a result.

Reading their results in Stockholm, and of course being familiar with much of the work, Meitner immediately reached the conclusion that the uranium atom had indeed fissioned, and was the first to publish a report concerning the possibility. Szilard, now in the United States, recognized that uranium fissioning into barium made a much more realistic candidate for a chain reaction than his proposed beryllium-helium reaction. He immediately contacted two other expatriate Hungarian physicists, Eugene Wigner and Edward Teller. The military potential that had led Szilard to keep his patent secret was now grimly apparent. The three physicists visited Albert Einstein, perhaps the only scientist in the world with the power to influence policy, and persuaded Einstein to write a carefully worded letter to President Franklin Roosevelt apprising him of the situation.

Roosevelt was sufficiently impressed that he took the initial steps that were to culminate in the Manhattan Project, the multiyear, two-billion-dollar effort that would produce the first atomic bomb.

Despite the plethora of scientific talent at America's disposal, all those involved realized that their German counterparts included not only Hahn but Werner Heisenberg, undoubtedly one of the most brilliant physicists in the world. Although the actual story is still not completely known, none of the German scientists who worked on the German atom bomb project were Nazi sympathizers, and as a result the project never received top priority from Hitler. Hahn and Heisenberg were taken into custody by American forces at the end of the war in Europe, and it was while being interned in England that they were notified of the bombing of Hiroshima. Hahn felt personally responsible, and for a while considered suicide. Like many (but by no means all) of the scientists associated with nuclear fission, he became a staunch opponent of nuclear weapons, refusing to cooperate with a West German project to manufacture them.

THE CREATION OF RADIOACTIVE ISOTOPES

When your mother is unquestionably history's greatest female scientist, and your father was also a brilliant scientist before his untimely death in a traffic accident, your chances of becoming a great scientist yourself must be substantially better than average. So it was with Irene Curie, the elder daughter of Pierre and Marie Curie. Educated privately and steeped in the scientific tradition, Irene served as her mother's assistant while Marie Curie continued her lifelong investigation of radioactivity.

During this period, Irene met Frederic Joliot, a chemist of outstanding promise who had been specially selected to become one of Marie Curie's assistants. They were married in 1926. Of course, it is standard practice for the wife to take the husband's name, but as Pierre and Marie Curie had no sons, Frederic Joliot also took his wife's name as a measure of his reluctance to let the great name of Curie die. Irene and Frederic Joliot-Curie lived together and worked together, as had Marie and Pierre Curie before them.

After Marie Curie became too ill to continue her work, Irene succeeded to her position. Over the next few years, the Joliot-Curies compiled a maddening record of scientific near-misses. In 1932, they were on the verge of discovering the neutron, but the English physicist James Chadwick beat them to it. In 1933, they almost discovered the positron whose existence had been predicted by Dirac, but the American physicist Carl Anderson found it while studying cosmic rays.

In 1934, however, they hit paydirt—or rather, they created it. They were investigating the results of colliding alpha particles with light elements such as aluminum. This experiment, similar to the ones conducted by Ernest Rutherford in his studies of the nucleus, knocked protons out of aluminum nuclei. They discovered that even when they ceased bombarding the target with alpha particles, the nuclei still emitted a form of radiation, even though they were no longer emitting protons.

The Joliot-Curies soon realized that in bombarding the aluminum, they had created phosphorus, but not the phosphorus that is usually found in nature. Natural phosphorus is not radioactive; the phosphorus created in the experiments was. The Joliot-Curies had artificially created isotopes that could not be found in nature.

This discovery had both theoretical and practical implications. Prior to these experiments, it had been thought that radioactivity was a phenomenon confined to heavy elements such as uranium and thorium. Now it was realized that any element could be radioactive if one only prepared the correct isotope. Since then, more than a thousand different radioactive isotopes have been prepared. These have been used in medicine, agriculture, industry, and scientific research. Radioactive tracers can be prepared, and are much safer to use than those radioactive elements that occur naturally. As a result, far greater use has been made of artificially produced radioactive isotopes than of naturally occurring radioactive materials. In 1935, the Joliot-Curies joined the Curies as the second husband-and-wife team to receive a Nobel Prize.

During World War II, the Joliot-Curies rendered a service of inestimable value to the Allied cause. Realizing the importance of heavy water to the construction of an atomic bomb, they smuggled the entire French supply of heavy water out of the country. This may have been one of the reasons why the Nazi atomic bomb project, which was headed by the brilliant physicist Werner Heisenberg, never really got off the ground.

During the war Frederic became an avowed Communist, and Irene worked for many organizations with Communist affiliations. As a result, when Irene applied in 1954 for membership in the American Chemical Society, the McCarthy-inspired paranoia that was sweeping the nation resulted in her application being rejected.

Inside the Atom

Originally, Max Planck's quantum hypothesis was viewed as more of a mathematical trick than anything else—it resolved the ultraviolet catastrophe, but it was not viewed as anything more significant. When Einstein used it to explain photoelectricity, it was clear that the quantum hypothesis was more than simply a mathematical trick; it was able to explain phenomena that other theories could not.

But the quantum theory did much more than explain physical phenomena. It had a profound impact on one of the most basic of philosophical questions: what is the nature of reality? What quantum theory had to say about this question puzzled the greatest physicists of the era—and a century later, many of the most fundamental questions raised by quantum theory are still unresolved. Quantum theory has forever changed the world, from the devices it has enabled to be constructed to the questions it has raised that transcend the world of physics.

THE UNCERTAINTY PRINCIPLE

Scientists tend to view the world either visually or symbolically, and there have been brilliant scientists of each type. However, as physics probed ever-deeper into the subatomic world in the first few decades of the twentieth century, it became harder and harder to visualize the phenomena that were occurring. As a result, some physicists, of which Werner Heisenberg was one, preferred to attempt to treat the subatomic world through symbolic representation alone.

Heisenberg had the good fortune of being one of Niels Bohr's assistants, and as a result was thoroughly familiar with Bohr's "solar system" model of the atom. However, Bohr's model was running into certain theoretical difficulties, and several physicists were trying to resolve them. One was Erwin Schrödinger, whose solution entailed treating the subatomic world as consisting of waves, rather than particles. Heisenberg adopted a different approach. He devised a mathematical system consisting of quantities known as matrices, which could be manipulated in such a fashion as to generate known experimental results. Both Schrödinger's and Heisenberg's approaches worked, in the sense that they accounted for more phenomena than Bohr's atomic model. In fact, the two theories were later shown to be equivalent, generating the same results using different ideas.

In 1927, Heisenberg was to make the discovery that would not only win him a Nobel Prize, but would forever change the philosophical landscape. In the late eighteenth century the French mathematician Pierre Laplace made a statement that would characterize scientific determinism. Laplace stated that, if one knew the position and momentum of every object in the Universe, one could calculate exactly where every object would be at all future times. Heisenberg's uncertainty principle states that it is impossible to know exactly where *anything* is and where it is going at any given moment. These difficulties do not really manifest themselves in the macroscopic world—if someone throws a snowball at you, you can usually extrapolate the future position of the snowball and maneuver to get out of the way. On the other hand, if both you and the snowball are the size of electrons, you're going to have a problem figuring out which way to move, because you will not know where the snowball will go.

Heisenberg's uncertainty principle is sometimes erroneously interpreted as an *inability* on the part of fallible humans to measure phenomena sufficiently accurately. Rather, it is a statement about the limitations of knowledge, and is a direct consequence of the quantum-mechanical view of the world. As a fundamental part of quantum mechanics, the uncertainty principle has real-world ramifications for the construction of such everyday items as lasers and computers. Even more profoundly, it has banished the simple cause-and-effect view of the Universe that had been unquestioned since the Greek philosophers first enunciated it. Heisenberg stated one of the consequences of the uncertainty principle as follows:

> In the experiments about atomic events we have to do with things and facts, with phenomena that are just as real as any phenomena in daily life. But the atoms or the elementary particles themselves are not as real; they form a world of potentialities or possibilities rather than one of things or facts. . . . Atoms are not things.

If atoms are not things, what are they? Seventy-five years after Heisenberg's revelation, physicists—and philosophers—are still struggling with this question.

The description of Werner Heisenberg given above undoubtedly presents a picture of an intellectual struggling with deep and profound questions. It would be hard to reconcile this image with the title of a song by the Rolling Stones: "Street-Fighting Man." Yet, at the end of World War I, Werner Heisenberg was indeed a street-fighting man, engaging in pitched battles with Communists in the streets of Munich after the collapse of the German government following the war. Perhaps this can be regarded as a youthful indiscretion, as Heisenberg was only a teenager at the time.

COMPLEMENTARITY AND THE QUANTUM VIEW OF REALITY

From the moment we first began to engage in intellectual speculation, we have wondered about the nature of reality. This question has preoccupied every generation of thinkers, beginning with the first recorded works of the great Greek philosophers. It is a question that could be posed by a child, but a satisfactory answer has eluded all who have tackled it.

Ever since Isaac Newton incorporated mathematics as an essential part of a description of natural phenomena, it has generally been easier for a theoretician to sit down with pencil and paper and derive mathematical consequences, than for an experimenter to devise and carry out a successful experiment. As a result, there is sometimes the feeling that mathematics is merely a convenient language to describe phenomena, but it does not give us an intuitive insight into the nature of the phenomena.

One such classic example was Dalton's atomic theory. As a result of the assumption that matter consisted of atoms, it was easy for chemists to work out the bookkeeping of chemical reactions. A chemist could write down the equations of a chemical reaction and predict in advance which substances, and in what quantity, would be produced. However, for more than a century after Dalton, individual atoms could not conclusively be shown to exist. To some scientists, Dalton's theory might have been simply a convenient mathematical description, a formalization that would tell you what would happen without necessarily telling you the way things actually were.

The rise of quantum mechanics in the first quarter of the twentieth century produced a more sophisticated version of this problem. The mathematical formulation of quantum mechanics viewed electrons as probability waves rather than actual things. To some, this was merely a convenient mathematical fiction. After all, a bunch of electrons constituted an electric current, which ran motors, and the electrons in the outer shells of atoms reacted chemically to form real substances such as water. How could they not be real themselves?

Albert Einstein championed the "reality" point of view, and to illustrate the problem, he and physicists Boris Podolsky and Nathan Rosen devised a thought experiment, which has come to be known as the EPR experiment. According to the laws of physics, it is possible for two photons (call them A and B) to be emitted so that the total spin (a quantum-mechanical property) of the two photons is known. Quantum mechanics dictated that the spin of a photon is not known until it was measured, and that the act of measuring this photon is part of the input that determines the result of the measurement. Consequently, once the spin of photon A has been determined, the spin of photon B could be calculated.

Einstein, Podolsky, and Rosen objected to this. Before the measurements, neither spin is known. Suppose two groups of experimenters, light-years apart, set out to measure the spins of these photons. If the spin of photon A is measured, and seconds later the spin of photon B is measured, quantum mechanics predicted that photon B would "know" the result of the measurement on photon A, even though there would not be enough time for a signal from photon A to reach photon B and tell photon B what its spin should be!

According to Einstein, this left two choices. One could accept the so-called Copenhagen interpretation of quantum mechanics, due primarily to Niels Bohr, that photon B knows what happened to photon A even without a signal passing between them. Alternatively, one could believe that there is a deeper reality, manifested in some physical property as yet unfound and unmeasured, which would supply the solution to the above dilemma without having to believe in photons possessing faster-than-light intuition. Einstein died holding firmly to this latter view—the physical property as yet unfound and unmeasured would be called a "hidden variable."

In 1964, the Irish physicist John Bell showed that if hidden variables existed, then experiments could be performed in which mathematical relationships known as "Bell inequalities" would manifest themselves. Within a few decades, hardware would be constructed enabling these experiments to be performed. Since then, sophisticated experiments with ultrafast lasers, which were not possible during Einstein's lifetime, have all but shut the door on Einstein's belief. Recent versions of what are called "quantum eraser" experiments have left only the narrowest of loopholes for Einstein's views to squeeze through. Maybe it really is true that the Universe is not only stranger than we imagine—it is stranger than we are capable of imagining.

QUARKS

Science sometimes appears to be an endless series of revelations, as prevailing explanations are succeeded by new ones. Often this is the result of a three-step procedure of collecting data, organizing the data into a pattern, and constructing a theory that explains the reason for the pattern. Examples of this from chemistry would be the search for new elements, Mendeleev's construction of the periodic table, and Bohr's description of the atom.

Man has been searching for the ultimate constituents of matter since the Greeks first conjectured them to be air, water, earth, and fire. After atoms were revealed to have structure in the first quarter of the twentieth century, it appeared that the question might have been solved: atoms consisted of a cloud of electrons orbiting a nucleus of protons and neutrons. However, it was quickly apparent that this model of matter was incomplete.

In order to probe structure, it is necessary to use more and more energy the deeper one goes. To pry electrons away from an atom does not require much energy; this is the level at which ordinary chemical reactions take place. To break up the nucleus requires substantially more energy. By the middle of the century, particle accelerators capable of delivering ever greater amounts of energy had been built for precisely this purpose.

The results were fascinating from an experimental standpoint but disturbing from a theoretical one. More than a hundred different particles were found, a virtual "zoo" waiting for a modern Mendeleev to come along and organize them. In the early 1960s, both Murray Gell-Mann and Yuval

Ne'eman arrived at an organizational chart called the Eightfold Way. In almost a repeat of the Mendeleev story, there was a gap in the chart that demanded the existence of a particle with specific properties; that particle was quickly discovered. The Eightfold Way was the periodic table of particles, but was there an underlying idea that could explain it?

In the late 1950s, Robert Hofstadter had conducted probes of protons and neutrons at the highest energy levels then available. He discovered that protons and neutrons were not hard, point-like particles, but rather appeared to have some sort of internal structure. This made it appear possible that neutrons and protons could be composite particles themselves.

Gell-Mann and George Zweig, another physicist from Caltech, independently devised a system of particles that would account for this. These particles would combine in sets of three to make protons and neutrons, and other combinations could account for the other particles. Gell-Mann called them "quarks," based on a line in James Joyce's *Finnegan's Wake*, "Three quarks for Muster Mark." The appellation stuck.

Decades of experimentation have confirmed that there are three families of quarks. The first family, the up and down quarks, make up the protons and neutrons of ordinary matter. The other two families are the charm and strange quark, and the top and bottom quark, and they make up the more exotic particles that are produced in high-energy processes, either by particle accelerators or as cosmic rays. Accompanying the three families of quarks are three families of leptons. Each family consists of an electron and a neutrino. This arrangement of particles comprises what is called the Standard Model. The last of the particles in the Standard Model, the Higgs boson (which is responsible for particles having mass) was predicted to exist by the Australian physicist Peter Higgs in 1964. Nearly half a century later, in 2012, the Large Hadron Collider confirmed the existence of the Higgs boson, and in 2013 Higgs was awarded the Nobel Prize. Better late than never!

It used to be possible for a Mendeleev or a Rutherford to reveal the most intimate secrets of nature in a small laboratory, in a reasonably short time, for not too much money. This happy situation seems to have come to an end. To discover the top quark required the efforts of thousands of scientists and technicians, working for more than a decade, and an expenditure of billions of dollars. If there is an even deeper level of organization than the Standard Model, how many people will have to work on it, how long will it take, and can we afford it?

CHAPTER 5

Forces and Energy

One of the great triumphs of science is the Standard Model of physics, which has taken centuries to formulate. It tells us what the components of matter are, and the forces that enable matter to interact with other matter. Work, in physics, is a measure of how much of an interaction has taken place, and energy is the capacity for doing work. A good measure of technological and scientific progress is how much work we are able to do, and how well we are able to harness and store energy to do that work.

The Industrial Revolution

For most of our history, work was done either by physical labor—human or animal—or by flowing water. Fire had been harnessed for cooking, heating, and a few industrial processes such as the smelting of ore or the making of glass, but it wasn't until the invention of the steam engine that heat was utilized to substitute for tasks that previously had required physical labor. Steam engines enabled heavy lifting and rapid transportation, and their importance to society prompted scientists to examine more closely the nature of heat.

THE LAWS OF THERMODYNAMICS

In the summer of 1847 William Thomson, a young British scientist, was vacationing in the Alps. On a walk one day from Chamonix to Mont Blanc, he encountered a couple so eccentric they could only be British—a man carrying an enormous thermometer, accompanied by a woman in a carriage.

Thomson, who later became known as Lord Kelvin, engaged the pair in conversation. The man was James Prescott Joule, the woman his wife, and they were in the Alps on their honeymoon. Joule had devoted a substantial portion of his life to establishing the fact that, when water fell 778 feet, its temperature rose 1 degree Fahrenheit. Britain, however, is notoriously deficient in waterfalls, and now that Joule was in the Alps, he certainly did not intend to let a little thing like a honeymoon stand between him and scientific truth.

A new viewpoint had been arising in physics during the early portion of the nineteenth century: the idea that all forms of energy were convertible into one another. Mechanical energy, chemical energy, and heat energy were not different entities, but different manifestations of the phenomenon of energy. James Joule, a brewer by trade, devoted himself to the establishment of the equivalence between mechanical work and heat energy. These experiments involved very small temperature differences and were not spectacular, and Joule's results were originally rejected by journals and the Royal Society. He finally managed to get them published in a Manchester newspaper, which might have published them because Joule's brother was the paper's music critic. Joule's results led to the first law of thermodynamics, which states that energy cannot be created nor destroyed, but only changed from one form to another.

Some twenty years before Joule, a French military engineer named Nicolas Carnot had been interested in improving the efficiency of steam engines. The steam engine developed by James Watt was efficient, as steam engines went, but nonetheless still wasted about 95 percent of the heat used in running the engine. Carnot investigated this phenomenon and discovered a truly unexpected result: it would be impossible to devise a perfectly efficient engine, and the maximum efficiency was a simple mathematical expression of the temperatures involved in running the engine. This was Carnot's only publication, and it remained buried until it was resurrected a quarter of a century later by William Thomson (Lord Kelvin), just one year after his chance meeting with Joule in the Swiss Alps.

Carnot's work was the foundation of the second law of thermodynamics. This law exists in several forms, one of which is Carnot's statement concerning the maximum theoretical efficiency of engines. Another formulation of the second law, due to Rudolf Clausius, can be understood in terms of a natural direction for thermodynamic processes: a cube of ice placed in a glass of hot water will melt and lower the temperature of the water, but a glass of warm water will never spontaneously separate into hot water and ice.

The Austrian physicist Ludwig Boltzmann discovered an altogether different formulation of the second law of thermodynamics in terms of probability: systems are more likely to proceed from ordered to disordered states (which explains why a clean room tends to get dirty, but dirty rooms do not tend to become clean). The first and second laws of thermodynamics seem to appear in so many diverse environments that they have become part of our collective "understanding" of life: the first law says you can't win, and the second law says you can't even hope to break even.

Carnot, Joule, and Boltzmann came at thermodynamics from three different directions: the practical (Carnot), the experimental (Joule), and the theoretical (Boltzmann). They were linked not only by their interest in thermodynamics, but by difficult situations bordering on the tragic. Carnot died of cholera when he was only thirty-six years old. Joule suffered from poor health and a childhood spinal injury all his life and, though the son of a wealthy brewer, became impoverished in his later years. Boltzmann was bipolar, and suffered from depression so severe that despite a rich circle of family, friends, and admiring students (among whom was Lise Meitner, who helped discover nuclear fission), he committed suicide because he feared his theories would never be accepted. Sadly and ironically, his work was recognized and acclaimed shortly after his death.

Electricity and Magnetism

Six hundred years before the birth of Christ, Thales of Miletus discovered that he could create sparks of static electricity by rubbing amber with fur. In fact, the word "electricity" comes from *elektron*, the Greek word for amber. Four hundred years later, the Chinese were using lodestones to create primitive magnetic compasses by rubbing an iron needle with a lodestone, which would cause the needle to be magnetized and point to indicate true north. But it took almost another two thousand years before science reached a full understanding of these phenomena, and how they were related.

THE NATURE OF ELECTRICITY

There are basically two types of discoveries in science. Some discoveries are carefully planned, the results of well-designed experiments or lengthy observation of a particular phenomenon. Some discoveries, however, happen almost entirely by chance. Perhaps it is not surprising that most of the amusing anecdotes surrounding a discovery relate to those that happened by chance.

One of the most well-known discoveries in science occurred as a result of chance to the Italian anatomist Luigi Galvani. One day in 1771, he noticed that some dissected frog legs twitched excitedly when struck by an electric spark. The anecdote surrounding this particular incident concerns a possible reason that the dissected frog legs were there in the first place: they were to be one of the main items of Galvani's evening dinner.

At first, Galvani was interested but not too surprised, because it was well-known that muscles would twitch when struck by an electric spark. Such demonstrations were used as part of an evening's entertainment. Then Galvani realized that he could use this to obtain confirmation of Benjamin Franklin's discovery that lightning was electrical in nature. He then attached the frog muscles to brass hooks outside his window so that they rested against some iron bars. As Galvani expected, the muscles twitched when a lightning bolt split the sky.

However, they also twitched on a perfectly clear day, as long as they made contact with two different metals. Electricity was certainly appearing from somewhere, but was it being generated by the metals or by the muscle? Galvani reached the conclusion that the electricity was being generated by the muscle, and called this type of electricity "animal electricity."

Alessandro Volta was a friend of Galvani; in fact, Galvani sent Volta copies of his papers. Like many other scientists of the period, Volta was interested in electricity. Early in his career, he invented a device called an electrophorus, which could be used for storing electrical charge. The principle behind the electrophorus is still used in today's electrical condensers.

Upon reading of Galvani's experiments with frog muscles, Volta decided to try to resolve the problem of whether the current Galvani had observed came from the muscle or from the metals. His experiment was simple; he decided to use only the metals and eliminate the frog muscles altogether. He immediately discovered that an electric current was produced, and naturally concluded that the difference in metals was the source of that current. Galvani disagreed, and a furious debate ensued.

In 1800, Volta provided the definitive proof by constructing devices that would produce electricity by means of two different types of metals. He constructed three types of discs: copper, zinc, and cardboard dipped in salt. He then stacked these discs in the following order, reading from the bottom up: copper, zinc, cardboard, copper, zinc, cardboard, and so on. When a wire was connected from the bottom disc to the top one, a sustainable electric current would pass through the wire. This device was called a "Voltaic pile" by Volta's contemporaries. We call it a battery.

Both Galvani and Volta are immortalized in language—we speak of being galvanized into action, and the unit of electromotive potential is the volt. However, the conquests of Napoleon had a drastically different effect on the lives of the two scientists. Galvani lost his job when he refused to swear allegiance to a government installed by Napoleon. He then lost his professorship, and died shortly thereafter. Volta felt that science was more important than politics, and visited Napoleon in France to demonstrate his experiments. Even after Napoleon fell, the resilient Volta prospered under the succeeding government.

THE LAWS OF ELECTROSTATICS AND ELECTRODYNAMICS

Scientific investigation is a product of the intellect. Political events are often products of everything but the intellect, yet scientific investigation is often greatly affected by political events. At the end of the eighteenth century, the key political event was the French Revolution, which had a vastly different impact on the lives of Charles Coulomb and André Ampère.

Coulomb was living in Paris when the Revolution began, and promptly moved to the provincial town of Blois where he could work without being disturbed by the uproar. He had already made his initial contributions to science by then, which included having invented an extremely delicate torsion balance, which he used to measure the effects of electrical charge. In *Principia*, Newton gave the defining property of a force as its ability to cause acceleration, and Coulomb was able to show that electricity was a force very much like gravity: its strength varied inversely as the square of the distance between charged objects. Although gravitation is always an attractive force, electricity can be both attractive or repulsive depending on whether the two charged objects have like or unlike charge. This result is known as Coulomb's law.

The effect of the French Revolution on André Ampère was much more drastic. A child prodigy who had mastered advanced mathematics by the age of twelve, his world was shaken when his father, who was a city official, was guillotined. Tragedy was again to dog Ampère's footsteps, as his wife died after a few years of marriage. Nonetheless, he was able to pursue a successful career as a professor of mathematics.

In 1820, word of Danish physicist Hans Oersted's experiment, which showed that a compass needle could be deflected by a nearby wire carrying an electrical charge, reached the French Academy of Sciences. Ampère, who was also interested in physics, decided to analyze Oersted's experiment in order to predict in which direction the compass needle would move. Within a week he had formulated the right-hand rule known to all physics students. Ampère was the first to attempt a mathematical treatment of electrical and magnetic phenomena, which would reach fruition with Maxwell's analysis some forty years later.

Georg Ohm was affected by the French Revolution only in the most indirect fashion. Raised in a poor family by a father who was a mechanic, Ohm's highest aspiration was to receive an appointment to teach in a university. After the French Revolution, the French mathematician Jean Fourier had accompanied Napoleon on an expedition to Egypt, where he had devised a novel theory on the nature of heat flow. Ohm decided to apply Fourier's ideas on heat flow to the flow of electric current in a wire. He was able to show that the amount of current was directly proportional to the potential difference and inversely proportional to the amount of resistance in the wire.

This relationship, known as Ohm's law, was not initially recognized as an achievement that would merit a university position. As a result, Ohm was forced to endure both professional disappointment and financial hardship. Gradually, his work began to be more appreciated outside his native Germany, and he was made a member of the Royal Society in England. Eventually, Ohm's law was recognized (even in Germany) as one of the key results in the theory of electrostatics, and he became a professor at the University of Munich.

Coulomb, Ohm, and Ampère each received one of the highest forms of praise that the scientific community can bestow, having units of measurement named in their honor. Electrical charge is measured in coulombs, resistance in ohms, and current in amperes.

There is a curious link connecting Coulomb and Ohm—Henry Cavendish, one of the most eccentric scientists who ever lived. Cavendish completely avoided women, even to the extent of ordering his female servants to stay out of his way or else they would be fired. He never left his house except to attend meetings of the Royal Society. He never changed his style of dress, which was extremely old-fashioned for that era (although everything from that era seems old-fashioned to us). There is only one portrait of Cavendish extant, and it looks as if he was wearing clothes that were a century out of date. Cavendish was a brilliant scientist, but his notebooks and manuscripts remained hidden for nearly a century after his death. When they were discovered, it was found that he had anticipated both Coulomb's law and Ohm's law, the latter by almost fifty years.

THE PRINCIPLE OF ELECTROMAGNETIC INDUCTION

Shortly after 4:00 p.m. on November 9, 1965, at a Queenston, Ontario, power station, an automatic control device that regulated and directed the flow of electric current abruptly failed. As a result, a circuit breaker that should have closed remained open, and a huge surge of electricity suddenly poured into the grid supplying power to the northeastern United States. From Boston to Rochester, generator safety switches automatically tripped, taking the generators offline so that they would not be damaged. In a domino effect, huge portions of the network simply "shorted out" to prevent the system from being permanently crippled, much as a blown fuse in a house will prevent an electricity overload from starting a dangerous fire. With the exception of buildings such as hospitals, which had generators for emergency situations, the electrical lifeblood of the entire Eastern seaboard was suddenly cut off. Lights went out, elevators stopped, and subways ceased to move. As night fell, New York was plunged into a darkness deeper than any it had experienced for more than a century.

The almost complete dependence of civilization on electrical power is a consequence of one of the most important experiments in the history of science. In 1831, Michael Faraday demonstrated that, if a magnet were moved through a coil of wire, an electric current would flow in the wire. This is known as the principle of electromagnetic induction, and is the basis for the production of electricity.

Faraday's experiment was the result of a simple but brilliant idea. In 1820, Oersted had shown that turning an electric current on and off near a compass would cause the compass needle to move. Faraday reasoned that if an electric current could affect a magnet, perhaps a magnet could be made to affect an electric current.

The vast increase in the wealth of our society, made possible by cheap and widely available electric power, occurred because Faraday's discovery has enabled us to tap into the power of both the gravitational force and the Sun. The Sun's heat evaporates water from the ocean. It rises, cools, and falls as rain or snow at high altitudes. Gravitational force causes water to run downhill, where we take advantage of the falling water to rotate large magnets in machines known as dynamos. This induces an electrical current, which can be efficiently transported far from the source by transmission lines. When we plug a device using an electric motor into an outlet, the electric current causes magnets to move, and it is this motion that enables the appliance to operate. The Sun's heat eventually evaporates the water that powered the dynamos, and the cycle starts again.

Faraday also possessed a keen intuition regarding the electrical and magnetic forces he was studying. Many of the great scientific advances are made possible by new ways of conceptualizing phenomena. Faraday visualized electricity and magnetism as consisting of lines of force permeating space, with stronger forces creating a greater concentration of lines in a particular region. This method of visualizing electricity and magnetism led to the idea of a field, a type of mathematical description that occupies a central position in physics.

Michael Faraday's parents apprenticed him to a bookbinder in London, so that he might learn a trade. This turned out to be an ideal situation for Faraday, as it gave him plenty of opportunity to read, especially about the subject that most appealed to him, science. In 1812 the famous chemist, Sir Humphry Davy, gave a series of lectures on chemistry for the general public. Faraday attended these lectures, and took copious notes. He then wrote the equivalent of a fan letter to Davy, who was so impressed that he gave Faraday a job as his assistant. Though lacking the formal education possessed by most scientists, Faraday was quick to make his mark in both chemistry and physics. He was made a member of the Royal Society, and when Davy retired, Faraday was given Davy's professorship.

At the age of forty-eight, Faraday joined a long line of scientists, including both Davy and Isaac Newton, who had suffered a nervous breakdown. Some of these nervous breakdowns were undoubtedly psychological in nature, such as Boltzmann's. Quite possibly, though, Faraday's may have been brought on by exposure to toxic chemicals, as the chemists of the day had no idea of the hazards of the chemicals with which they were in daily contact.

THE FIELD THEORY OF ELECTROMAGNETISM

By the end of the eighteenth century, scientific investigation of both electric and magnetic phenomena was well under way. From a strictly qualitative standpoint, the two phenomena have things in common: both appear to come in two varieties (positive and negative charge, north and south magnetic poles), and obey the law that opposites (positive–negative, north–south) attract and likes (positive–positive, north–north) repel. Indeed, when Coulomb discovered late in the eighteenth century that both electricity and magnetism obey inverse-square laws similar to the one obeyed by gravity, the search for theories of electricity and magnetism analogous to Newton's theory of gravitation had begun in earnest.

Half a century after Coulomb's discoveries, two significant experiments revealed that electricity and magnetism must be related in some fashion. Oersted had shown that a compass needle would deflect in the presence of an electric current, and Faraday had demonstrated that motion inside a magnetic field would induce an electric current. Coulomb's discoveries, and the Oersted and Faraday experiments, spawned numerous attempts to create satisfactory mathematical theories of electricity and magnetism. Each of the competing theories had its strengths, but no one theory successfully described the full range of phenomena.

Into this arena came James Clerk Maxwell, probably the most brilliant theoretical physicist to live between Newton and Einstein. Maxwell had already made remarkable contributions to various areas of physics. He had demonstrated that the rings of Saturn could not be completely solid, as the tidal interactions would tend to tear them into smaller, asteroid-type bodies, a prediction that was visually confirmed in the 1980s by the Voyager satellites. He had also done significant work on the theories of gases, heat, and statistical mechanics.

His crowning achievement, however, was the unification of electric and magnetic phenomena—a set of four vector equations relating electric and magnetic fields, which were to become known as Maxwell's laws. The hallmark of a great mathematical theory in science is that mathematical deductions can predict new physical phenomena. In Maxwell's theory, electromagnetic waves could be shown to travel with a speed equal to the measured speed of light, causing Maxwell to predict that light would be shown to be an electromagnetic phenomenon. In addition, the oscillation rate of the waves could vary significantly, leading Maxwell to predict the existence of types of electromagnetic waves, beyond the infrared and the ultraviolet, which had not yet been detected. Shortly after Maxwell's death, Heinrich Hertz detected the radio waves that Maxwell's theory had predicted.

Newton described gravity in terms of a force exerted instantaneously between two objects, whereas Maxwell described electromagnetism in terms of vector fields throughout space. Maxwell's success in creating a field theory of electromagnetism sparked an upswing of interest in creating field theories to describe other phenomena. Maxwell himself tried unsuccessfully to develop a field theory of gravity, but this was not achieved until the work of Einstein. Interestingly enough, Einstein's work superseded much of classical physics, but Maxwell's laws remained inviolate even in an Einsteinian Universe. Einstein then began *his* unsuccessful quest for a unified field theory, a quest that, under the name "theory of everything," continues to this day.

SUPERCONDUCTIVITY

While it is easy to produce extreme heat, it is not so easy to produce extreme cold, and extreme cold is what is needed in order to liquefy many gases. Heike Kamerlingh Onnes broke through the cold barrier in 1906, liquefying hydrogen at about 20 degrees above absolute zero, and again in 1908, liquefying helium at only 4 degrees above absolute zero. Helium was the last of the gases to be liquefied, and Kamerlingh Onnes decided to investigate the properties of materials at extremely cold temperatures.

His most startling discovery was that, at extremely low temperatures, metals lost their resistance to electricity, a phenomenon he named superconductivity. Resistance causes electric current to dissipate as heat, and it enabled the development of such useful gadgets as the electric heater and the electric toaster. However, resistance has a major disadvantage; it creates loss of current when current is transmitted through cables and wires. The mere existence of superconductivity hinted that it might be possible to transmit electricity long distances without loss.

Kamerlingh Onnes also discovered that each metal had a characteristic temperature at which it became superconducting. He called this the transition temperature, and the goal was to find high-temperature superconductors. Progress in this direction was agonizingly slow. There was no theoretical understanding of superconductivity, and for a long time the highest known transition temperature was in the vicinity of 20 degrees above absolute zero.

In 1957, John Bardeen, Leon Cooper, and Robert Schrieffer developed a theory of superconductivity that appeared to explain the phenomenon. Although it was a valuable theoretical insight, it had no immediate practical impact. The materials with the highest transition temperatures were niobium compounds, and niobium is relatively hard to obtain. Investigation into superconductivity was moribund for decades.

In 1986, Karl Müller and his student Georg Bednorz, working at the IBM laboratories in Zurich, came up with a startling discovery. They discovered a material with a transition temperature 50 percent higher than the previous record—an astonishing 35 degrees above absolute zero. Even more startling was the fact that the superconducting material was a ceramic oxide—not even a metal.

The next few months saw transition temperature records falling on an almost daily basis. One immediate goal was to push the transition temperature past 77 degrees above absolute zero, as that is the point at which nitrogen becomes liquid. At low temperatures, liquid helium is needed, and this is quite expensive, whereas liquid nitrogen is commercially available and inexpensive.

As a result, superconducting ceramic oxides are currently being exploited in a variety of devices, such as magnetic resonance imaging (MRI) scanners.

At present, history is repeating itself. There is still no universally agreed-on theoretical understanding of the ceramic superconductors, and to make matters worse, ceramics cannot easily be shaped into wires or cables, the primary means of transmitting electricity. The highest transition temperature is currently a subject of controversy, as an as-yet-unconfirmed experiment has pushed the threshold of superconductivity to close to room temperature. But room temperature superconductivity has tremendous potential technological and economic benefits, and efforts to find such materials are ongoing and will doubtless continue until they are either found or shown to be nonexistent.

The name John Bardeen is not well known, yet he holds two Nobel Prizes. The second of his Nobel Prizes was for the theory of superconductivity that has been described above. The first (which he shared with Walter Brattain and William Shockley) was for the invention of the transistor, a device that has revolutionized the second half of the twentieth century. Winning two Nobel Prizes puts one in a league with such titans of science as Albert Einstein and Marie Curie.

Light

Light is not just a phenomenon; it is a metaphor. "Let there be light," according to the Bible, was the first thing God did after creating the heavens and the earth. Light is seen as good, as contrasted with the evil represented by darkness.

Probably no question in science has created more controversy over a longer period of time than the nature of light. Greek and medieval philosophers alike puzzled over it, alternating between theories that light was a substance and that light was a wave, a vibration in a surrounding medium.

THE DOUBLE-SLIT EXPERIMENT

The debate found two great physicists in the seventeenth century to champion each viewpoint—perhaps the first of the really great scientific debates—as it takes great concepts and great scientists to make for a really epic scientific debate. Isaac Newton, when he wasn't busy with mathematics, mechanics, or gravitation, found time to invent the science of optics. Newton believed that light consisted of tiny particles, as that would explain the nature of reflection and refraction. The eminent Dutch scientist, Christiaan Huygens, however, championed the point of view that light was a wave.

What are some of the characteristics of waves? Sound, one classic example of a wave, can go around corners. Light doesn't. Water waves, another obvious type of wave, can interfere with each other. When two water waves collide, the resulting wave can be either stronger or weaker than the original waves, stronger if the high points of both waves reinforce each other, and weaker if the high points of one wave coincide with the low points of the other wave.

Such was the almost universal reverence in which Newton was held that few efforts were made to either validate or dispel the wave theory of light for over a century. The individual who would finally perform the definitive experiment was Thomas Young, a child prodigy who could read by age two and who could speak twelve languages by the time he was an adult. In addition to being a child prodigy, fortune had favored Young in other respects, as he was born into a well-to-do family. After a brilliant performance as a student at Cambridge, Young decided to study medicine.

Young was extremely interested in diseases and conditions of the eye. While still a medical student, he discovered how the shape of the eye changes as it focuses. Shortly after, he correctly diagnosed the cause of astigmatism, a visual fuzziness caused by irregularities in the curvature of the cornea.

Young's fascination with the eye led him to begin investigations into color vision and the nature of light. In 1802, he performed the experiment that was to show once and for all that light was a wave phenomenon. Young cut two parallel slits into a piece of cardboard and shone a light through the slits onto a darkened background. If light were a substance, there should have been bright patches directly opposite the slits, with the intensity diminishing to either side of the slits. This is what one would expect if one were to spray paint or some other substance at the slits.

What Young observed, however, were alternating bright bands of light interspersed with totally dark regions. This is the classic signature of wave interference. The bright bands occurred where the "high points" of the light waves coincided, the dark regions where the "high points" of one light wave were canceled out by the "low points" of the other.

Young's double-slit experiment was unquestionably important to establish the wave nature of light, but its significance goes substantially beyond that. With the advent of quantum mechanics, the wave or substance nature of many subatomic phenomena became a subject of debate. Double-slit experiments have been employed in different guises to answer many of the subtle questions that occur in this most perplexing area of physics.

Thomas Young was a polymath whose accomplishments extended into many of the realms of science, and even beyond. In addition to the achievements that have already been noted, he constructed a theory of color vision, observing that in order to be able to see all colors, it was only necessary to be able to see red, green, and blue. He made significant contributions to the theory of materials; Young's modulus is still one of the fundamental parameters used to describe the elasticity of a substance. Young was also an Egyptologist of note, and was the first individual to make progress toward deciphering Egyptian hieroglyphics.

THE INVISIBLE ELECTROMAGNETIC SPECTRUM

There are two complementary pleasures that an experiment in science can elicit. The first is to discover a phenomenon that you had no idea existed. The second is to discover one that you predicted should exist, but has not yet been observed.

Friedrich Wilhelm Herschel was born in Hannover, Germany. At the time of his birth, Hannover was a British possession, and Herschel moved to England at the age of nineteen and changed his name to William. A successful career as a musician made him financially secure, and he decided to pursue his interest in astronomy. Since a good telescope cost far too much for him to purchase, he decided to grind his own lenses. He had returned to Hannover to bring his sister Caroline to England, and Caroline had acquired William's addiction to astronomy.

Anything having to do with lenses and optics interested Herschel. One day in 1800 he decided to use thermometers to measure whether there was any difference in the temperature associated with the differing colors of the spectrum. He used a prism to spread sunlight out into the familiar ROY G BIV colors of the rainbow: red, orange, yellow, and so on. In the process, he accidentally left a thermometer beyond the red of the spectrum. Examining the setup sometime later, he was surprised to find that the thermometer outside the visible colors was also registering an increase in temperature. The obvious conclusion was that there existed a color beyond red that was invisible to the naked eye. This color is now called infrared.

One year later, Johann Ritter, a German physicist, was experimenting with photography. It had been discovered that silver chloride turned black in the presence of light, and that blue light was more efficient than red light in causing this reaction. Ritter discovered that there was a light beyond violet that was even more efficient at turning silver chloride black. This color is now known as ultraviolet.

More than eighty years later, Maxwell had explained that visible, infrared, and ultraviolet light were all examples of electromagnetic radiation. In theory, electromagnetic waves of any length should be detectable. Heinrich Hertz had constructed an electrical circuit that generated an oscillating spark across a gap between two metal balls. Maxwell's equations predicted that such an oscillating spark should produce electromagnetic waves. Using a simple detector consisting of a loop of wire terminating in two small metal balls separated by an air gap, Hertz was able to pick up the wave by moving around the room; at various points a spark would jump across the gap in the detector loop at the moment another spark was generated in the primary circuit.

Hertz calculated that the wavelength of the electromagnetic radiation generated by the spark was 66 centimeters, a million times the wavelength of visible light. This experiment served to confirm Maxwell's theories, but it was to have an even more far-reaching effect. Within fifteen years, Guglielmo Marconi was able to devise a practical way to communicate with these "Hertzian waves." Nowadays, "Hertzian waves" are much better known as radio waves, and Marconi's invention was, of course, the radio. You may well have found yourself slightly annoyed by a modern version of Hertz's experiment when you accidentally generate a spark while your radio is on, as your radio will then act as a detector for the spark by emitting static.

William Herschel was unquestionably the outstanding observational astronomer of his day, but of course his astronomy was limited to visible light. He would have been extremely pleased to know that the infrared radiation he discovered has been exploited in observational astronomy, as have ultraviolet radiation and radio waves. In fact, observational astronomy is currently conducted throughout the entire electromagnetic spectrum. Many of the great discoveries in astronomy over the past three decades have been made by telescopes built to observe electromagnetic radiation outside the range of visible light. WMAP, the Wilkinson Microwave Anisotropy Probe, placed the age of the Universe at 13.8 billion years and its composition at 27 percent ordinary matter and 63 percent dark matter. SOFIA, the Stratospheric Observatory for Infrared Astronomy, discovered that helium hydride was the first molecule to be formed in the Universe.

THE DOPPLER EFFECT

The phrase "the Doppler effect" sounds like the title for a thriller or science-fiction movie. It is a phenomenon familiar to all of us, and it lies at the heart of a variety of everyday—and not-so-everyday—devices.

Christian Doppler was an Austrian physicist in the first half of the nineteenth century. It was not a good time to be an Austrian physicist, and Doppler had a difficult time obtaining an academic position. As many did at that time, he made plans to immigrate to the United States, the land of opportunity. Just as he was getting ready to leave, he received an offer of a professorship in Prague. As a result, Doppler stayed in Europe.

The Doppler effect was originally discovered in conjunction with sound waves. Early in the seventeenth century, it had been noticed that sound failed to travel through a vacuum, but would travel through air and water. Such behavior was characteristic of waves, and in the eighteenth century, Marin Mersenne had computed the speed of sound in air to an accuracy of 10 percent. Newton had actually been the first to attempt a mathematical analysis of sound, and by the early nineteenth century the behavior of sound waves as generated in organ pipes or vibrating strings was fairly well understood.

Doppler, along with many others, had observed that the pitch of a sound, which is the aural perception of its frequency as a wave, varies if the sound is being generated from a moving source. The sound seems more highly pitched as the moving source approaches the listener, and more deeply pitched as it moves away from the listener. This phenomenon can easily be observed by listening to the whistle of an approaching train or, since trains are not as common as they used to be, the siren of an approaching police car.

Doppler correctly reasoned that the moving source should impart its motion to the waves. As the moving source approaches, the wave crests reach the listener more rapidly, thus increasing the frequency and raising the pitch. As the moving source departs, the wave crests take longer to reach the listener, with the opposite effect.

Once he had worked out the equations, Doppler conceived of one of the most charming experiments in the history of science to test his conclusions. He managed to get several trumpeters to sit on top of a flat car that was being pulled by a locomotive. The trumpeters were instructed to play a particular note, and the train would proceed at a set speed either toward Doppler or away from him. Two days of experiments confirmed his deductions.

Although the Doppler effect was first applied to sound waves, it can be used for light waves as well. In everyday life, the Doppler effect is used in "speed guns," which can determine not only the speed of a thrown fastball, but also the speed of a moving car. More importantly—at least from the standpoint of this book—is that the Doppler effect is responsible for one of the most profound deductions in the history of science. In the 1920s, Edwin Hubble observed that the light from the majority of galaxies has been "red shifted"—that is, the frequency of the emitted light waves had decreased. From this he deduced that most galaxies are moving away from the Earth, and was able to derive a relationship between their speed of recession and their distance from us. This has not only helped us determine the size and age of the Universe, it also was an important step on the road to realizing that the Universe began in a big bang.

As of this writing, the Doppler effect is one of several ways to search for planets outside the solar system. The gravitational effect of large planets on a star causes that star to wobble slightly, and the light from that star is Doppler-shifted by the wobble. This can be picked up by sensitive instruments. As of this writing, almost five thousand exoplanets have been discovered, nearly one thousand of them through the use of the Doppler effect.

THE NONEXISTENCE OF THE ETHER

Unexpected results from experiments force scientists to think more deeply about the validity of their present theories, and to construct new explanations that fit the unexpected results. The question of whether light consisted of particles or waves appeared to have been settled in 1803 when Thomas Young performed his famous double-slit experiment showing that light produces interference patterns. Since waves in a liquid also produce interference patterns, this seemed to clinch the case for the wave theory.

Water waves move through water, and sound waves move through air (sound waves cannot be transmitted through a vacuum), so the obvious question was: what was the medium through which light waves were transmitted? The French physicist Augustin Fresnel, who performed extensive investigations of the wave theory of light, named this substance "ether." Although no one had ever been able to isolate and study ether, the prevailing view through the middle of the nineteenth century was that waves could not exist without something to travel in, and therefore the ether must exist.

In 1887 the American physicists Albert Michelson and Edward Morley decided to measure how the Earth moved through the ether by using Michelson's newly invented interferometer, a sensitive measuring device based on interference patterns. They assumed the Earth was stationary, and measured the speed of light in two perpendicular directions. To everyone's surprise, the two speeds were identical, implying that the Earth was standing still. This result was obviously wrong, since it had been known from the time of Galileo that the Earth moved through the Universe.

This totally unexpected result forced scientists to think again about the nature of light. Two European physicists, George Fitzgerald and Hendrik Lorentz, found that they could resolve the problem under the apparently absurd assumption that objects contracted as they moved more rapidly. This would explain the Michelson–Morley result, as the measuring device would contract in the direction the Earth was moving, and thus the speed of light would appear to be the same no matter how the Earth moved.

In 1905, Albert Einstein published his theory of special relativity, which explained how the hypothesis offered by Fitzgerald and Lorentz was warranted. Einstein assumed that the laws of physics were the same in any two systems moving at a constant velocity relative to each other. This forced the speed of light to be the same in any two such systems. In the same year, known to physicists as Einstein's "Miracle Year," he also explained the photoelectric effect by assuming that light behaved as a particle. Special relativity showed that there was no need for the ether to exist as a frame of reference, and the particle theory of light eliminated the need for a medium through which light waves would travel. The current thinking of physicists is that light is both wave and particle; and the resolution of this apparent paradox is one of the foundations of modern physics.

Michelson's interferometer has proved to be one of the most important measuring devices ever invented. Indeed, one of the staples of modern radio astronomy is the technique of VLBI (Very Long Baseline Interferometry), in which radio telescopes at opposite ends of the Earth are linked by computer to resolve the structure of extremely distant objects. Interferometers are similar to telescopes in that the larger the lens, the more powerful the tool, and VLBI enables astronomers to create interferometers whose "lenses" are effectively as large as the Earth itself.

Other Forces

By the last decade of the nineteenth century, physicists had identified two major forces—gravity and electromagnetism. There were similarities between the two—both could easily be seen to be manifested in everyday life, and both obeyed an inverse-square law; objects three times as distant from one another exerted forces that were only one-ninth (1 divided by 3^2) as strong.

However, in the manner of a novel of the supernatural, unseen forces were at work, and much of the twentieth century would be devoted to understanding how these unseen forces help to shape not only the world we inhabit, but the Universe itself.

THE DISCOVERY OF X-RAYS

Every scientist dreams of discovering a new phenomenon. New phenomena open up unsuspected possibilities, changing the ways we think and the things we can do. Luck undoubtedly plays a role in the discovery of new phenomena, but, as has been said, "Fortune favors the prepared mind." It may be added that fortune favors the observant eye as well.

In this case, the observant eye belonged to Wilhelm Roentgen, head of the department of physics at the University of Wurzberg in Bavaria. The fact that Roentgen had attained this position reveals that he was already a respected scientist. Roentgen had experienced rebelliousness as a youth, having once been expelled from school for making fun of a teacher, but after finally obtaining a degree in mechanical engineering he switched to the study of physics, and had established a solid reputation. On November 5, 1895, he was experimenting with cathode rays and the luminescence they induce in certain materials.

So that he could more clearly see the faint luminescence, he darkened the room and covered the cathode ray tube (which emitted light) with black cardboard. He turned on the cathode ray tube, and a flash of light some distance from the tube caught his eye. Searching for the source of the light, he discovered a piece of paper coated with barium platinocyanide glowing in the dark. Roentgen was puzzled, because he was aware that cathode rays, the type of ray produced by the cathode ray tube, were incapable of penetrating the black cardboard surrounding the tube.

He turned off the tube, and the sheet of paper darkened. Turning the tube on again caused the paper to glow again. Obviously something emanating from the tube was causing the paper to glow. Then Roentgen did something that indicated he was a first-class experimenter: while holding the paper, he walked into the next room, and darkened it. With the cathode ray tube still on in the adjacent room, the paper continued to glow. Whatever was emanating from the cathode ray tube could actually pass through walls. You can imagine Roentgen's excitement—nothing even remotely like this had ever been observed before.

Roentgen now found himself facing the classic scientist's dilemma: to rush findings into print to make sure of being first, or to nail down the phenomenon through rigorous experimentation. He chose the latter course, burning the midnight oil for almost seven weeks. When he submitted his first paper on what he called X-rays (X is often the mathematician's choice to describe something unknown), it was not merely a preliminary report, but included many of the important physical properties of X-rays. Later, when being questioned by a naive interviewer who asked Roentgen what he thought when he discovered X-rays, he replied, "I didn't think, I experimented."

On January 23, 1896, Roentgen delivered his first lecture on the new phenomenon. At the end of the lecture, he asked for a volunteer from the audience, and an octogenarian walked up to the stage. An X-ray photograph was taken of his hand, developed, and displayed to the audience, showing the bones clearly. It brought down the house. Four days after the news reached America, X-ray photographs were used to locate a bullet in a man's leg. Within a year, a thousand papers had been published on X-rays.

It is known today that X-rays are an extremely powerful, and potentially dangerous, form of electromagnetic radiation. Properly used, X-rays have made tremendous contributions to medicine and dentistry. Among the developments that would have undoubtedly surprised and delighted Roentgen is X-ray astronomy, a recent field that has revealed dramatic insights into some of the most energetic events occurring in the Universe.

So revolutionary were X-rays that, when the Nobel Prizes were established, Roentgen won the first such prize for physics. Roentgen was a scientist to the core, refusing to accept a title for his accomplishment from the King of Bavaria, and made no attempt to obtain what would have been an extremely lucrative patent for X-rays, an action that shocked the financially astute inventor, Thomas Edison. Unfortunately, this philanthropic decision came back to haunt Roentgen. After World War I, the subsequent inflation financially destroyed many Germans, including Roentgen, who died in near poverty.

THE DISCOVERY OF RADIOACTIVITY

When something is passed from father to son for several generations, the assumption is that what is being passed is either control of the family business or the old homestead. In the case of the Becquerels, however, it was the professorship of applied physics at the Museum of Natural History in Paris.

Antoine Becquerel's grandfather had fought side by side with Napoleon. After Napoleon's defeat at Waterloo, he switched from a military career to a scientific one, eventually becoming the first Becquerel to hold the aforementioned professorship of applied physics. Antoine Becquerel's father succeeded to this position, spending much of his time studying fluorescence and phosphorescence. These two phenomena occur when matter absorbs light at one wavelength and emits it at another; it is especially beautiful when a mineral absorbs invisible ultraviolet light and emits it in a different, and visible, color.

When Becquerel succeeded to the family business (physics) in 1892, he continued his father's investigations of these subjects. Then, in 1895, Wilhelm Roentgen's startling discovery of X-rays commanded Becquerel's (and almost every other scientist's) attention. Becquerel wondered whether any of the fluorescent materials he had been studying might be emitting X-rays. This was a reasonable conjecture, as Roentgen had discovered X-rays because they caused certain materials to fluoresce.

In February 1896, Becquerel wrapped photographic film in black paper, placed a crystal of potassium uranium sulfate on top of it, and placed the entire assemblage in direct sunlight. It was known that sunlight caused potassium uranium sulfate to fluoresce. If the fluorescence contained X-rays, the X-rays might penetrate the black paper, which visible and ultraviolet light could not, and fog the photographic film. Sure enough, when Becquerel developed the film, it was fogged. Becquerel reached the natural conclusion that fluorescence involved X-rays.

He was understandably anxious to continue his experiments, but Nature intervened with a series of cloudy days. Becquerel prepared his experiment of film, black paper, and crystal, but the weather would not cooperate by delivering the sunlight necessary to induce fluorescence. Overcome by impatience, he one day decided to develop the film anyway, perhaps hoping that some faint fogging would show up; this would at least enable him to correlate the degree of fogging with the amount of sunlight.

On developing the film, Becquerel was astounded to discover that the plate was as strongly fogged as when the crystal had been exposed to a substantial amount of direct sunlight. With the instincts of a top-flight experimenter, Becquerel realized that the fogging was being produced by the crystal rather than the sunlight. He experimented with other uranium compounds, and even with metallic uranium itself. The results were always the same, enabling Becquerel to realize that it was not a chemical phenomenon he was encountering, but an atomic one related to the atomic structure of uranium itself.

Becquerel had discovered the phenomenon of radioactivity, one of several turn-of-the-century discoveries that were to revolutionize physics. As a result of Becquerel's discovery, scientists realized that vast sources of energy were obtainable not only through chemical and mechanical processes, but through atomic ones as well.

As all great discoveries do, Becquerel's discovery of radioactivity stimulated many other scientists to investigate the phenomenon. Two other scientists who were so motivated were Pierre Curie and his wife, Marie. The Curies soon discovered that thorium, radium, and polonium were also radioactive. In 1908, the Nobel Prize for physics was awarded jointly to Becquerel and the Curies. Marie Curie's life was to be inextricably intertwined with radioactivity—she not only studied it extensively and named the phenomenon, but her death was due to its cumulative effects.

SEMICONDUCTORS AND TRANSISTORS

As the experimenters of the seventeenth and eighteenth century discovered, although both gravity and electricity obey an inverse square law, there are many significant differences between them. While gravity acts equally on all substances, different substances can have vastly different electrical properties. A copper wire and a glass rod fall at the same rate, but electrical current passes easily through a copper wire and not through a glass rod.

A copper wire is called an electrical conductor; the term is thought to have originated with that most daring of electrical experimenters, Benjamin Franklin. The reason that the copper wire conducts electricity so well was not known until the structure of the atom was discovered early in the twentieth century. Electrical current is simply the flow of electrons, much as a current of water is the flow of drops of water. In a metal atom, such as copper, the electrons in the outer shell of the atom are easily induced to leave the atom by the presence of an electric field. In a molecule of glass, the electrons are so tightly secured that it requires an electrical field of incredible strength to induce them to move. Materials such as glass are called insulators.

In a copper wire, the flow of current can proceed in either direction along the wire; this behavior is characteristic of a conductor. However, physicists had discovered certain substances during the nineteenth century, such as silicon, germanium, and gallium, having the unusual property of permitting current flow in one direction only. These substances are called semiconductors.

In 1929, the Swiss physicist Felix Bloch studied these substances, and formulated a theory involving the population of different bands by electrical charge. According to Bloch, under normal circumstances the electrons of a semiconductor's atoms were trapped in what he called valence bands. However, if the semiconductor received the right amount of energy from an outside source, electrons can jump from the valence bands to what Bloch called forbidden bands, and this enabled current to pass.

In 1945, three physicists at Bell Telephone Laboratory, John Bardeen, Walter Brattain, and William Shockley, created the first transistor. The name "transistor" was derived from the fact that the semiconductor could be manipulated to *trans*fer current across a re*sistor*. A transistor could both rectify current (allow it to pass in one direction only), but even more importantly, it could amplify current—if a small electrical charge was passed to the transistor, a larger electrical charge would be emitted. These properties allowed the much simpler, stabler, and cheaper transistor to perform functions that had previously been done by the more complicated, erratic, and expensive vacuum tubes.

This is not a book about invention, but the importance of the transistor to science, as well as to everyday life, simply cannot be overestimated. It not only revolutionized consumer electronics, it made possible many of the great achievements of science and engineering in the last half of the twentieth century. The first transistor was an ungainly affair consisting of solid-state devices soldered to a plate with protruding wires. By the end of the twentieth century, millions of transistors would be combined in devices called microprocessors, which are the "brains" behind practically every piece of advanced electronics produced today.

The importance of the transistor was quickly recognized, as the 1956 Nobel Prize in physics went to Bardeen, Brattain, and Shockley. Bardeen was later to share a second Nobel Prize for his share in developing the BCS (Bardeen–Cooper–Schrieffer) theory of superconductivity. Shockley was to later become infamous for taking the controversial position that racial differences on IQ tests might be the result of genetic factors, as opposed to environmental ones.

THE WEAK AND STRONG FORCES

From the standpoint of physics, the Universe is described by particles and the forces that act on those particles. There are four forces. A mathematical description of gravity was first given by Isaac Newton, and the field theory of electromagnetism by James Clerk Maxwell. Gravity and electromagnetism are long-range forces; two particles placed at opposite ends of the Universe will attract each other gravitationally, and if they have electric charges they will attract or repel each other electrically.

As the twentieth century dawned, it became clear that atoms themselves had structure, and a great deal of effort has been expended into probing this structure. It became apparent that each atom consisted of a nucleus of protons and neutrons, surrounded by a cloud of electrons. There was an obvious problem associated with this model of the atom. If the protons in the nucleus had a positive electrical charge, why didn't the nucleus simply fly apart as the protons all repelled each other?

In 1935 Hideki Yukawa, a Japanese physicist, came up with a solution that seems almost obvious in retrospect. There had to be some force stronger than electrical repulsion holding the protons together. Another characteristic this "strong force" must display was that it acted only over very short distances; if it acted over longer distances, it would certainly have already been detected.

According to our understanding, forces act by exchanging particles whose mass increases as the distance over which the forces act decreases. The photon is the particle whose exchange makes electromagnetism possible. Photons act over infinite distances, and have zero mass. Yukawa was able to calculate the approximate mass of the particle whose exchange would make possible the strong force. In 1947, experiments confirmed Yukawa's prediction, for which he won the Nobel Prize in 1949.

The fourth force is the weak force, which is responsible for the phenomenon of radioactivity. It, too, is a subatomic force, acting over very short distances. Although theories have been constructed to hypothesize a fifth force, no experiments have ever confirmed its existence—at least, to the satisfaction of the scientific community.

Many physicists operate under the assumption that, at the instant of the big bang, there existed only one force in the Universe. As the Universe cooled, other forces came into existence, much as water and ice form as steam cools. Showing that two forces are actually aspects of the same phenomenon is known as unification. James Clerk Maxwell's theory of electromagnetism was a unification of electricity and magnetism.

Throughout the 1960s, independent work by Sheldon Glashow, Abdus Salam, and Steven Weinberg suggested that at higher temperatures, electromagnetism and the weak force would be unified into an "electroweak" force. Just as Yukawa had predicted the existence of a particle to carry the strong force, Glashow, Salam, and Weinberg predicted the existence of three distinct particles to carry the electroweak force. In order to observe these particles, it was necessary to create extremely high temperatures, which could only be done in massive particle accelerators. Using a radical new particle accelerator, a team of physicists at CERN (the European Organization for Nuclear Research) discovered all of these particles in 1983.

Now that electroweak unification has been achieved, the search is on to unify all four forces. A theory that unifies the electroweak force with the strong force is known as a GUT (grand unified theory), and a theory that unifies all four forces is known as a TOE (theory of everything). However, in order to verify a TOE, using current technology it would require building a particle accelerator almost as large as the Universe! This raises the intriguing, though highly nonscientific, question: was our Universe the result of an attempt to prove a TOE in some other Universe?

CHAPTER 6

Life

Life is probably the biggest mystery in the Universe. Its unanswered questions—such as "How did life begin?" and "Is there life elsewhere in the Universe?"—are among the most tantalizing questions that have ever been asked. But within the next few decades science will almost certainly answer the former, and there is a good chance it can answer the latter as well.

Of course, these questions have been asked long before there was such a thing as science. Because these questions are so profound, and there is such a need for answers, nonscientific answers have occupied a central position in philosophy and religion. When science has come up with answers to these questions, conflicts have arisen between science and religion—and these conflicts continue today.

Varieties of Life

Even before written records, man undoubtedly noticed the other forms of life that could be seen by the naked eye—plants, insects, animals, fish, and birds. Life existed in abundant varieties, and an obvious first move was to categorize the various forms of living organisms.

CLASSIFICATION OF LIVING ORGANISMS

It is impossible to truly appreciate Aristotle unless one either reads a biography or at least examines an encyclopedia. Only then will it become clear that he was one of history's most brilliant thinkers. He is perhaps best known as a philosopher, but he is responsible for creating and systematizing the development of logic, and was a first-rate scientist as well. Aristotle's works were lost for several centuries following the fall of Rome, and it is perhaps not coincidental that these were the darkest of the Dark Ages. When his works were rediscovered, the power of his thought was so astonishing that for centuries he was the unquestioned authority on practically everything.

Aristotle spent a great deal of time and effort observing the different animal species. In so doing he displayed the attributes of both a naturalist and a scientist. He was especially interested in species that lived in the sea, and recorded the ability of the torpedo fish (also known as the electric eel) to stun its prey, even though he knew nothing of electricity. Noticing that dolphins gave birth the same way as mammals, he correctly classified them as mammals rather than fish. In fact, Aristotle's primary contribution to biology was that he devised a hierarchical classification scheme for over five hundred different animal species.

In so doing, he may have been one of the first scientists to have an intimation of the ideas of evolution. Aristotle's hierarchies progressed from simpler to more complex forms, and he was aware that this may have occurred as the result of some progressive change. However, it would have been impossible for him to stumble upon the mechanism of natural selection, and he was far too good a scientist to have endorsed such a theory without being able to devise a way to support it.

Nearly two thousand years after Aristotle, the explorations of the seventeenth and eighteenth centuries discovered many species unknown to Aristotle, stretching his classification scheme past the breaking point. The importance of a useful classification scheme cannot be overestimated, as there are many instances in science where such schemes and great breakthroughs have gone hand in hand.

Carl Linnaeus, a Swede who combined the roles of naturalist and biologist in the same way as Aristotle, devised the system that we still use today. For each species of living thing, he wrote down a clear description that pointed out how that species differed from other, similar species. He also popularized the binomial nomenclature that is still used, in which each species is given a generic name and a specific name. Modern man, for instance, is *homo sapiens*—homo denotes the group ("man"), and sapiens the specific characteristic ("knowing").

Linnaeus was not content merely to classify; he organized as well. Species with many similar traits were organized into a group called a genus. Similar genera were organized into an order, and similar orders into a class. The basic idea of this system is still being used, although it has been expanded somewhat to accommodate not only the new animal and plant species that are continually being discovered, but life-forms such as bacteria and viruses of which Linnaeus was entirely unaware.

Linnaeus lived in an era in which religion, philosophy, and science were not as separate as they are today, and his life reflects this. He believed that he had been directed by God to oversee this project, and thought of people who did not agree with him as heretics. Despite these traits, reminiscent of a religious zealot, he was known to be a caring and inspirational teacher who trained many future scientists to continue his work.

THE DISCOVERY OF BACTERIA

If there is one quality associated with science in the public mind, it is doubtless intellectual brilliance. A person who is brilliant is sometimes described as "an Einstein," and to indicate that you don't need to be a genius to do something, we say, "It doesn't take a rocket scientist to . . ." We know that the achievements of Newton and Einstein transcend our abilities—even if we had been given Kepler's data or the results of the Michelson–Morley experiment, we wouldn't have come up with the theory of universal gravitation (in the former case) or the theory of special relativity (in the latter).

So it may be surprising that one of the most significant achievements in the history of science could have been performed by anyone with good eyesight who had happened to look in the right place at the right time. It was actually a Dutch grocer, Anton van Leeuwenhoek, who had the curiosity to examine a drop of rainwater under the lens of a microscope, and it was Leeuwenhoek who thus became the first person to observe the world of bacteria—the invisible zoo.

Leeuwenhoek was an extremely competent observer who thought nothing of observing the same object as much as a hundred times to make sure that he had assimilated all the details. His skill was such that he had become a "foreign correspondent" for London's Royal Society, communicating his observations along with detailed drawings. Nonetheless, his first description of the invisible zoo must have been hard for the members of the Royal Society to credit. "They stop, they stand still as 'twere upon a point, and then turn themselves around with that swiftness, as we see a top turn around, the circumference they make being no bigger than a fine grain of sand." The Royal Society promptly commissioned Robert Hooke, England's finest microscopist, to build a microscope sufficiently powerful so that Leeuwenhoek's findings could be confirmed—as, of course, they were.

What makes Leeuwenhoek's discovery so significant is its impact on medicine. The average life span nowadays is decades more than it was in Leeuwenhoek's era, and most of that increase is due to the ability to eliminate or cure disease. In Leeuwenhoek's day, disease was believed to be caused by evil spirits, and the attempts to cure disease often bordered on what we would call witchcraft today. Although it would be nearly two centuries until bacteria were actually associated with disease, without the observations Leeuwenhoek made, medicine would never have emerged from the Dark Ages.

So did Leeuwenhoek's discovery simply come as a result of being in the right place at the right time? Or was it the result of having the "highest tech" available in the form of the best microscopes of the era, which he himself had constructed? Undoubtedly, both were factors. However, microscopes were widely available at the time, and many observers were doing what Leeuwenhoek himself had also done—looking at the fine details of things too small to be seen with the naked eye, such as the hairs on a fly or the sting of a bee. It did not take much of an imagination to use a microscope to look more closely at such objects. Nevertheless, it must have taken a substantial amount of imagination to use a microscope to look more closely at a drop of ordinary rainwater—either imagination, or the curiosity that has always characterized the great scientists.

Leeuwenhoek was well aware that the quality of his microscopes gave him the edge as an observer that high tech usually does. Even though he was willing to allow others to use some of his lesser instruments, he once wrote, "I keep some for my own use and through these no men living hath looked save only myself." During his lifetime, he resisted even the pleas of the Royal Society. However, three months after his death, Isaac Newton received, on behalf of the Royal Society, a cabinet containing twenty-six of Leeuwenhoek's finest microscopes.

THE DISCOVERY OF ANAEROBIC ORGANISMS

It would probably not be a difficult task to write a book entitled *Louis Pasteur's Top 20 Contributions to Science.* Pasteur appears here yet again, as he is undoubtedly one of the seminal figures in the history of science. Here we examine one of Pasteur's lesser-known contributions, which would have made the career of almost any other scientist.

Pasteur was one of those rare individuals who view adversity as simply another challenge. After suffering a stroke in 1868 that almost killed him, and left him partially paralyzed, he still managed to produce some of his finest achievements.

By this time, the process of respiration was fairly well understood. It was known that plants use sunlight to manufacture carbohydrates from water and carbon dioxide, and produce oxygen as a waste product during this process. Animals breathe the oxygen and consume the carbohydrates, producing carbon dioxide as a waste product. This is the great cycle involving plants and animals, the two primary kingdoms of life. It was felt that all forms of animal life were oxygen-breathers.

One of Pasteur's major interests throughout his life was the process of fermentation. It had been Pasteur who discovered that fermentation was caused by microorganisms. He had also saved the French wine industry by recommending that wine be heated in order to sterilize the organisms that were responsible for souring the wine.

In 1872, Pasteur happened to observe that air inhibited the movements of the bacteria that were responsible for changing sugar solution into butyric acid. Pasteur, whose intuition was legendary, immediately realized he had discovered an extremely interesting phenomenon. Further investigation revealed that oxygen was the inhibiting factor. Pasteur coined the word "anaerobic" to describe bacteria whose actions are limited by air. Today we know that there are two types of anaerobic organisms—those that function poorly in air, and the "obligate anaerobes" that are killed by exposure to oxygen.

We have also found that anaerobic bacteria exist in environments where one might suspect life does not exist. Anaerobic bacteria can be found in extremely salty environments, as well as extremely hot ones. Anaerobic bacteria are often found in hot springs, and some strains can even survive at temperatures close to that of boiling water.

One of the most interesting discoveries of anaerobic bacteria occurred in 1977. John Corliss and Robert Ballard, aboard the submersible *Alvin*, discovered the first "black smokers" near the Galapagos Islands. These are deep-sea vents that belch out superheated streams of hydrogen sulfide. This hydrogen sulfide was the energy source for a strain of anaerobic bacteria, and the bacteria formed the base of a food chain that involved large, ornately gilled tube worms and giant clams. This thriving community was located so deep in the ocean that it was impossible for sunlight to penetrate. The "black smoker" communities are the first known assemblage of life that does not rely on energy produced by the Sun, and might conceivably be descendants of the original life-forms to inhabit Earth. It also prompted many to consider that similar forms of life might have arisen elsewhere in the Universe.

Anaerobic bacteria have come to play an important role in forensic science. One of the key developments in DNA analysis was the discovery by Kary Mullis of the polymerase chain reaction, or PCR, which can create large quantities of DNA from very small samples by successive doubling. Mullis's original idea worked extremely slowly, because the heat needed for each doubling cycle killed the organisms involved in PCR, and so more organisms had to be added at each cycle. A major improvement in making PCR practical was to use Taq bacteria, an extremely heat-resistant anaerobe that could survive without needing to be replaced. This speeded up PCR substantially, turning it into a formidable tool. It has also, in the period of the coronavirus pandemic, supplied a major tool for the rapid analysis of whether someone has COVID-19.

THE DISCOVERY OF VIRUSES

The story of viruses, like so many of the great discoveries in biology, is a story spanning generations and continents. It starts in 1885, when the brilliant French chemist Louis Pasteur developed a vaccine to protect against rabies. Pasteur, who had spent much of his career working to prove the germ theory of disease, was unable to observe the germ that caused rabies. One of Pasteur's characteristics was a profound intuition concerning scientific phenomena, and he concluded correctly that the germ that caused rabies was so small, his microscope was unable to detect it.

Several of Pasteur's assistants would go on to make noteworthy contributions of their own. One of these assistants was Charles Chamberland, who in the course of working with Pasteur developed a series of filters for isolating bacteria. These filters, or variations of them, became standard tools in bacteriological research.

Several years later Dmitri Ivanovsky, a Russian botanist, was working with tobacco mosaic disease, a disease that caused the leaves of tobacco plants to become mottled. He used filters to try to isolate the bacteria causing this disease, but no matter how small the pores of the filter, the infective agent of the disease was able to slip through them. Ivanovsky, whose intuition was a notch or two below Pasteur's, reached the conclusion that the filters were defective, and had simply been unable to trap the bacteria.

A few years later Martinus Beijerinck, a Dutch botanist, performed a similar set of experiments relating to tobacco mosaic disease as had Ivanovsky. Beijerinck, however, came to a much different conclusion: the tobacco mosaic disease was caused by a living, though nonbacterial, agent that was small enough to pass through the pores of the filter. Moreover, whatever caused the disease was alive, as the infective agent would grow in a healthy plant and could be passed on to another healthy plant. Beijerinck named the unseen agent a "filterable virus," *virus* being the Latin word for "poison."

Beijerinck went on to demonstrate that numerous diseases, among which were polio, mumps, chickenpox, influenza, and the common cold, were caused by viruses. However, the nature of viruses remained unknown until Wendell Stanley, an American biochemist, performed an experiment whose results were completely unanticipated: he managed to crystallize the tobacco mosaic virus. This was not only an extraordinary experimental achievement, but one that raised a question still being debated today: are viruses alive? One of the criteria for life is that it be able to reproduce. Viruses can reproduce, but they cannot do so on their own. They must have a host cell whose reproductive machinery they can commandeer for this purpose.

The development of the electron microscope, which can magnify more powerfully than can optical microscopes, finally made it possible for scientists to see viruses and confirm Pasteur's intuition. Recent developments have made it possible to deduce the structure of viruses. A typical virus consists simply of a molecule of DNA surrounded by a protective protein coat. It is the smallness and simplicity of viruses that make them so difficult to defeat; there are only a limited number of strategies available to destroy a virus or prevent it from reproducing. This explains why, even though science (actually technology) can put a man on the moon, it can't cure the common cold—yet.

Had Wendell Stanley been born forty or fifty years later, he might have led the New England Patriots to their many Super Bowl victories, rather than Bill Belichick. As an undergraduate at Earlham College, Stanley played football and expected to be a football coach. While visiting the University of Illinois, he got involved in a conversation with a chemistry professor, and soon found himself more interested in chemical equations than football diagrams.

ANIMAL BEHAVIOR

Perhaps if science were conducted by computers and not by human beings, there would be no such thing as a pecking order among scientists. But science is conducted by human beings, and a pecking order has been in existence for more than a century.

Lord Kelvin, one of the foremost physicists of the nineteenth century, delineated the pecking order succinctly when he described naturalists and paleontologists as stamp collectors. If it could not be made into an equation, Kelvin didn't really consider it as science.

That prejudice exists even now. If one looks down the list of Nobel Prize winners in the category of physiology or medicine, the first true naturalists show up in 1973. Even then they were all grouped together, as if the Nobel Committee felt that by giving an award to one naturalist, they might as well give it to several so they wouldn't have to do it again soon. Each of these naturalists has a lifetime of achievement, but we will recount one of the career high points for each.

Karl von Frisch is best known for his brilliant investigation of communication in bees. One of the most important activities of a beehive is the discovery of sources of food. In order to expedite this, the colony sends out scout bees. If a scout bee finds an attractive food source, it brings pollen and nectar back to the hive. The next item on the agenda is to inform the hive of the location of the food source. Although the obvious thing would be for the scout to simply fly to the source, the scout's energy reserves are depleted and time is short. The scout communicates the location, in terms of distance and direction, by engaging in a complicated dance known as the "waggle" dance. The scout describes the distance by the time spent in waggling, and the direction relative to the sun by the angle the waggle is to vertical. Communicating such information is a pretty impressive accomplishment for insects; one wonders how it arose and got passed from generation to generation.

Konrad Lorenz is the author of a classic study on instincts in geese. When baby geese emerge from the egg, they recognize their mother not by sight but by a specific combination of sounds. Lorenz discovered that baby geese will assume that anything emitting these sounds is the mother, no matter how unlike a goose the author of those sounds might be. This procedure is known as imprinting. To demonstrate this, Lorenz himself emitted the appropriate sounds, and the baby geese imprinted on him as their mother. It is quite amusing to see a movie of Lorenz, in winter coat and galoshes and smoking a pipe, being followed by a number of baby geese.

Niko Tinbergen demonstrated the instinctive behavior of the stickleback, a type of fish, in defending its territory against intruding sticklebacks. Tinbergen discovered that the stickleback did not recognize other sticklebacks, but instead instinctively attacked anything with a characteristic patch of red on its underside (sticklebacks have red bellies). Male sticklebacks exposed to a stickleback-shaped dummy with a white patch on its underside showed very little reaction, but when they were exposed to circles or squares with red on the lower portion, they attacked viciously.

These discoveries have shed light on the complexities of animal behavior. In so doing, perhaps they can also illuminate some of our own.

One of the most intriguing studies in animal behavior would have satisfied even Lord Kelvin as to its scientific merits. The behavior of altruism in human beings can be understood from an emotional and intellectual standpoint—a man will risk his life to save a drowning child, even though he does not know the child. However, social insects such as wasps also display altruism; they will sacrifice their lives for others. William Hamilton performed a brilliant mathematical analysis to demonstrate that this behavior confirms Darwinian theory. He showed that the probability that an animal will display altruism increases with the number of genes the animal shares in common with the animal to which it is displaying the altruistic behavior.

The Science of Life

One of the reasons that the Dark Ages were dark is that scientific inquiry was largely inhibited by the religious belief that all the great questions had been answered—if not by Aristotle, then by the religious authorities. Most of what was done in terms of science took place in the Middle East, and communication between European and Middle Eastern cultures tended to take the form of military conflict. But, with the coming of the Middle Ages, people started to see how knowledge about the world around them could make life better in a measurable way, and the scientific inquiry that had started with the Greeks resumed.

Additionally, two major technological innovations greatly accelerated the advance of science—the telescope for astronomy, and the microscope for biology. These two devices are right up there with Gutenberg's invention of the printing press as the technological developments that have brought major advances to humanity.

THE DISCOVERY OF CELLS

If the personality of an individual is the result of his or her childhood experiences, Robert Hooke's childhood must have been a very unpleasant one. We do know that he was scarred from smallpox, and that he suffered what he deemed a humiliation when he went to Oxford, because he had to wait on tables to put himself through college. In any case, Hooke became a quarrelsome, jealous, and miserly individual. Despite his many achievements, it is even possible that, on balance, his net contribution to science was negative—he quarreled so violently with Isaac Newton that Newton had a nervous breakdown.

While at Oxford, Hooke acquired a justly deserved reputation for mechanical ingenuity. As a result, he began to work with Robert Boyle, who would eventually construct Boyle's law, a mathematical relationship between the temperature, volume, and pressure of a gas. It was Hooke who constructed an improved mechanical pump for Boyle to use in his investigations.

In 1662, Hooke's mechanical prowess earned him the position of "curator of experiments" for the Royal Society, the only paid position in the Society. A year later he was elected to the Royal Society. Instead of mellowing Hooke, these honors gave him a sword to wield against those he felt had wronged him, and he spent much of his career arguing that his ideas and devices had anticipated the achievements of others.

There was some justification in this, as he had planted seeds for several very important developments, but these seeds had not borne the significant fruit others would harvest. He had some of the basic ideas on the theory of gravitation that Newton eventually developed, and developed an erroneous wave theory of light that anticipated some of Huygens's thoughts. Hooke did make significant contributions, though, to astronomy, physics, and biology. In astronomy, he was one of the first individuals to discover double stars. In physics, he made an extensive study of springs, culminating in Hooke's law, which states that the restoring force on a stretched spring is proportional to the distance it is stretched from equilibrium. His investigation of springs eventually led to their use in watches and clocks.

It was in the field of biology, though, that Hooke's impact was greatest. He was England's greatest microscopist, using his mechanical ability to develop the compound microscope, with which he confirmed Leeuwenhoek's discovery of bacteria. In those days, microscopists would grab anything handy and put it under the microscope, and it was Hooke's good fortune one day to observe a thin sliver of dried cork. To his surprise, he observed a regular array of tiny rectangular pores. His father had been a clergyman, and the holes reminded him of the small rooms, known as cells, inhabited by monks in a monastery. Hooke named them "cells," and cells they remain to this day.

Hooke had been looking at a piece of dead cork, and the cells he described were actually what we now think of as live cells that had been drained of fluid. Cells are the basic building blocks of life, and occupy a position in biology similar to the one occupied by atoms in chemistry. Hooke made his discovery in 1665. The next year, London burned down in the Great Fire, and Hooke, occupied by its rebuilding and other aspects of science, never did any further work in microscopy.

Possibly because Hooke was such an unpleasant person, history has been reluctant to credit him with another significant contribution: he anticipated some aspects of Darwin's theory of evolution by almost two centuries. He used his microscope to examine fossils and, noting that no such creatures were still around, he wrote, "There may have been divers Species of things wholly destroyed and annihilated, and divers others changed and varied, for we find that there are some Kinds of Animals and Vegetables peculiar to certain Places and not to be found elsewhere." Although speculation is a useful adjunct to science, science requires proof, and Darwin (and Wallace) amassed the evidence to confirm Hooke's ideas.

THE CELL THEORY

Biology is in many respects a more complicated science than physics or chemistry. The structures with which it deals are more intricate, and the relationships between them sometimes exceedingly complex. As a result, biological theories are a little like biological objects; they tend to exhibit slow growth and arrive at maturity after a rather lengthy process. Such was the case of the cell theory.

When Robert Hooke first observed and named cells, he did so by looking at thin slices of dried cork. Since the cork was dead, the cell fluid was no longer present, and Hooke saw only the dried cell walls, not the activity that went on in the room. Over the next century and a half, microscopes improved to the point where biologists were able to observe living cells, and they noted that these cells were filled with fluid. Biologists of this period recognized that living things consisted of organs and tissues, but even though they could see cells within these structures, they did not regard cells as the building blocks.

Plant cells are easier to observe than animal cells because plant cells, unlike animal cells, have cell walls. The improving power of microscopes started to make itself felt in 1831, when the Scottish botanist Robert Brown first noticed a small dark body, which he named the nucleus, that appears in all plant cells.

Around this same time, a German lawyer named Matthias Schleiden was so unhappy in his profession that he attempted suicide. As a cure for depression, he took up the study of botany first as a hobby, and then as a profession. Rather than concentrating on plant classification, as most botanists of the time were doing, Schleiden preferred microscopic examination of plant tissues. In 1838, he stated his cell theory for plants: all plant tissues are made of cells. Moreover, Schleiden recognized the importance of the nucleus that Brown had discovered in the construction of new cells, although Schleiden erroneously thought that new plant cells budded from the nucleus.

One year later Theodor Schwann, a respected German physiologist, independently arrived at virtually the same conclusions for animal cells. Schwann also proposed that eggs are cells, and that all life starts as a single cell. Schleiden and Schwann had formulated the basics of cell theory, a theory that occupies roughly the same position in biology that the atomic theory does in chemistry.

Two decades later Rudolph Virchow, a respected German physiologist, was to unite the separate theories of Schleiden and Schwann under the pithy axiom "all cells arise from cells." This was the first theoretical broadside aimed at the theory of spontaneous generation, which Pasteur's experiments a few years later would completely demolish.

Nonetheless, even after Virchow, the process of the production of new cells still remained to be discovered. It would require a new generation, armed with even more powerful microscopes, to discover that cells actually divide by a process called "mitosis," and that it is this cell division that is responsible for the construction of new cells.

Virchow is one of many noteworthy scientists who have succumbed to the trap of falling in love with their own theory. He attempted to extend the cell theory to a theory of disease, hypothesizing that disease is caused when cells revolt against the organism to which they belong. Although one might regard forms of cancer in this light, diseases caused by germs obviously do not fall under its scope. As a result, Virchow refused to accept Pasteur's germ theory of disease, and Virchow also made it difficult for Robert Koch to obtain a position to continue the investigations in which he had attributed specific diseases to specific germs.

PHOTOSYNTHESIS

There are two great kingdoms of life on Earth: plants and animals. They exist in glorious harmony with one another, each quite literally existing by taking in the other's dirty laundry. Animals take in oxygen, the waste product of plants, and produce carbon dioxide via respiration, or breathing. Plants take in this carbon dioxide and produce both oxygen (which animals breathe) and starch (which animals eat) via photosynthesis.

The end of the eighteenth century witnessed dramatic developments in both politics and science, and many of the scientists of the time were involved in both. Inoculation as a preventive measure for smallpox was just becoming a recognized medical procedure, and Dutch physician Jan Ingenhousz lived in England long enough to become an expert in this area. As a result, he was called to Vienna to administer inoculations to the royal family, and eventually became the personal physician of the Empress Maria Theresa. He returned to England in 1779, where he was made a member of the Royal Society. At the same time, he began investigating the chemistry of plants.

Ingenhousz established the broad outlines of what is undoubtedly the most important chemical reaction on Earth, as it is the chemical reaction without which the animal kingdom would not exist. He established that plants took in carbon dioxide and gave off oxygen, but that this reaction would only take place in the presence of sunlight. To confirm this, Ingenhousz placed plants in a lightless environment, and established that the production of oxygen ceased. The term photosynthesis, which is used for this reaction, means "production in light."

The next major step in the investigation of photosynthesis was taken by Julius von Sachs, a German botanist. By the middle of the nineteenth century, the green pigment chlorophyll had been discovered and shown to be distributed throughout the green portion of the plant. Von Sachs showed that chlorophyll was confined to certain portions of the cell, which were later called chloroplasts, and it was within the chloroplasts that chlorophyll catalyzed the reaction that transformed carbon dioxide in the presence of sunlight to starch, in the process giving off oxygen as a waste product. This also explained why plants appear green. Sunlight consists of all colors of the spectrum, but chlorophyll does not use green light in the photosynthetic reaction, and the green light is thus reflected for us to see.

But what sequence of chemical reactions enabled chlorophyll to transform carbon dioxide into starch? Because the reactions could not be duplicated in a test tube, photosynthesis proved difficult to analyze until the American chemist Melvin Calvin began work in 1949. Calvin manufactured carbon dioxide with radioactive carbon and allowed plants only a few seconds to make use of the carbon dioxide. He then mashed up the plant cells and identified the individual components by a newly invented procedure known as paper chromatography. Those compounds containing radioactive carbon must have been manufactured very early in the photosynthetic process.

By 1957 Calvin had established the complete chain of chemical reactions by which photosynthesis is accomplished. Ingenhousz had established that plants take in carbon dioxide and give off oxygen, von Sachs had shown that this was done by chlorophyll inside the chloroplasts, and Calvin had determined the precise chemical reactions. From the discovery of photosynthesis to its complete description had taken almost two centuries.

The study of photosynthesis illustrates how critical technology can be to scientific advance. It took Calvin the better part of a decade to obtain his results, and he was using radioactive tracers and paper chromatography, the high-tech tools of his day. Modern technology can perform the same tasks thousands of times faster. Today, scientists use lasers that deliver light pulses lasting less than a billionth of a second to determine precisely how photosynthesis proceeds inside chlorophyll and bacteriorhodopsin, another light-sensitive molecule.

MITOSIS AND MEIOSIS

Every so often science stalls after an obviously great advance. This happened around 1840, after Schleiden and Schwann had put forth the theory of cells. The primary difficulty was that cells are composed mostly of transparent material, which made observation of the crucial details of cell processes extremely difficult.

However, just as there are times when progress stalls, there are also those times when a fortuitous discovery in another area turns out to be a catalyst for extremely rapid advance. The discovery by William Perkin of synthetic dyes in the early 1860s turned out to be a boon for cytologists, the scientists engaged in the investigation of cells.

Perkin, of course, thought of his discovery of synthetic dyes from the commercial standpoint of making colorfast fabrics. The cytologists, most notable Paul Ehrlich and Walther Flemming, soon learned that certain dyes were selectively absorbed by different portions of the cell. This technique was called staining.

Ehrlich and Flemming each used staining in a different fashion. Ehrlich, who was a student of Robert Koch, the father of bacteriology, used staining first in the identification of the germs responsible for different diseases. He then adapted the chemistry of dyes to the treatment of disease.

Flemming, however was a biologist, and he used the technique of staining to investigate processes within the cell. He coined the term "chromatin" from the Greek word for color, to describe the material within the cell that absorbed the dye.

Flemming dyed a large number of cells in growing tissue, and it was inevitable that some of these cells would be caught in different stages of the cell division cycle. As the process of cell division began, the chromatin organized itself into short, threadlike rods that Flemming called "chromosomes," a Greek term meaning "colored bodies." Whenever cell division took place, the appearance of these chromosomes was so characteristic that Flemming named the process "mitosis," from the Greek word for thread.

The key feature of mitosis is that the chromosomes double in number in the original cell. The chromosomes are then pulled apart to opposite ends of the cell, with half the chromosomes going to each end. The cell then divides into two cells. Because the chromosomes doubled in the original cell, each of the two resulting cells now has the same number of chromosomes as the original cell.

Mitosis ensures that, when cell division takes place, the two resulting cells are duplicates of each other. However, this is not the procedure that takes place in sexual reproduction, when the sperm cell and ovum join. The chromosomes from each cell do not double, but rather one from the sperm cell intertwines with the corresponding chromosome from the ovum cell in a process called "crossing over." The intertwined pair then splits up. This process, discovered by the Belgian cytologist Edouard Van Beneden, is termed "meiosis." The key aspect of meiosis is that a chromosome resulting from the union of sperm cell and ovum receives half of its genes from the sperm cell and half from the ovum. The discovery of meiosis provided the biological explanation for Mendel's laws of genetics, and is the reason why a baby may have her father's eyes, but her mother's nose.

Flemming actually realized that the process of mitosis did not occur in sexual reproduction, but was unable to document what actually happened. When Van Beneden worked out the details of meiosis, the two key parts of the puzzle for the understanding of genetics were present: what happened in meiosis, and Mendel's laws. Unfortunately, Mendel had died a few years previously, and his results had been ignored because statistics, with which Mendel had worked, was an unfamiliar tool to botanists. It wasn't until fifteen years later that Mendel's work was rediscovered, and its significance might not have been realized had the explanation of meiosis not existed.

How Life Changes

Ernst Haeckel, like Thomas Young, was a polymath—a naturalist who was, among other things, a physician, a philosopher, and a marine biologist. Haeckel was aware of the importance of marketing in science as well as in commerce, and he characterized one of his ideas with the catchy phrase "ontogeny recapitulates phylogeny." As Haeckel intended it, ontogeny—the development of an individual organism—parallels the development of the species, known as phylogeny.

Although this theory is no longer widely held, the idea that the individual life may in some way parallel something larger actually shows up in how the changing of an individual occurs in the two primary ways that species themselves change. The normal course of life is smooth—birth, childhood, adulthood, aging, and death—corresponding to evolution. But sometimes the life of an individual is cut short by disease or violence. The same thing happens on a larger scale, and a species becomes extinct through a more dramatic process.

THE THEORY OF EVOLUTION

Almost everyone has heard of Charles Darwin, but hardly anyone has heard of Alfred Wallace. Yet Charles Darwin and Alfred Wallace had essentially the same great idea, developed from similar experiences. If Wallace were alive today, he would not be surprised to learn that his name is nowhere near as widely known as Darwin's, because Wallace's life consisted of one piece of bad luck after another.

Darwin had been fortunate to come from a well-to-do family; his father was a wealthy doctor and his uncle, Josiah Wedgwood, was the head of the famous Wedgwood pottery manufacturers. As a result, Darwin could afford to take a position as unpaid naturalist on the H.M.S. *Beagle* when it sailed for the Galapagos Islands off Ecuador in 1831. Five years in this extraordinary locale gave Darwin many of the ideas that were later to become part of his epic book, *The Origin of Species*.

Alfred Wallace, however, had to earn a living, and so he became a surveyor. He later decided that he would rather try to make a living doing what he liked. In order to make a living as a naturalist he set sail in 1848 for South America to collect rare species. Four years of observing the profusion of life in the Amazon valley led Wallace to many of the same ideas that had occurred to Darwin. Unfortunately, after he had assembled his collection, he departed for England on a boat carrying a load of resin, a highly flammable substance. With typical Wallace luck, the boat caught fire, and the results of his four years of collecting were totally destroyed.

The parallels between their two lives were to continue. Both arrived back from South America with ideas about how species came about, but both initially could not conceive of a mechanism to generate new species. After returning to England, however, each happened to read Thomas Malthus's *Essay on the Principle of Population*, in which Malthus observed that populations multiply faster than food resources. It then occurred to both Darwin and Wallace that the species best adapted to its environment would obtain more of the resources necessary to support life. Thus was born the theory of natural selection, the mechanism by which evolution occurs.

All this occurred in 1838 to Darwin, who wrote up a thirty-five-page draft, which he expanded in 1842 to several hundred pages. Recognizing that it would be an intellectual bombshell, he arranged to have it published posthumously, possibly so that he could avoid the conflict he felt certain would occur. However, his other researches made him a well-known naturalist, and so when Wallace wrote up his own observations on evolution in 1858, he submitted a "preprint" to Darwin for comment.

Darwin was thunderstruck. As he put it, "I never saw a more striking coincidence; if Wallace had my MS. sketch written out in 1842, he could not have made a better short abstract!" Darwin now faced a dilemma, but friends of the two men arranged for both Darwin and Wallace to read their papers before a meeting of the Linnean Society in London. Hardly anyone noticed. However, Darwin's plans for a posthumous publication of his theory were no longer feasible. A year later, *The Origin of Species* was published. Ironically, Darwin, who would rather have avoided the fuss, would forever be the name associated with evolution.

Both Wallace and Darwin were passionate naturalists who from childhood had preferred traipsing through the countryside observing and collecting to more scholarly activities. Darwin had gone to Cambridge, but preferred collecting beetles to attending class (Wallace had evidenced similar preferences). As Darwin put it, "I will give a proof of my zeal: one day, on tearing off some old bark, I saw two rare beetles, and seized one in each hand; then I saw a third and new kind, which I could not bear to lose, so that I popped the one which I held in my right hand into my mouth." Wallace could undoubtedly have related similar tales.

THE DEATH OF THE DINOSAURS

The first dinosaur fossil was discovered in 1822 in England by Mary Ann Mantell. Twenty years later the British fossil hunter Richard Owen coined the term "dinosaur" from the Greek words meaning "terrible lizard." In 1854, Owen prepared an exhibit of dinosaurs for the Crystal Palace in London. This exhibit captured the public's fancy, beginning a love affair that shows no sign of abating, as the box office receipts for *Jurassic Park* and its sequels clearly demonstrate.

The first dinosaurs appeared approximately 225 million years ago, and were the dominant life-form on Earth for over 100 million years. At the end of the Cretaceous period, some 65 million years ago, the dinosaurs abruptly vanished. It was one of the great mysteries of paleontology: what killed the dinosaurs?

Numerous theories were advanced to explain their disappearance. One possibility was that the newly evolved mammals, smaller and faster, ate the eggs of the dinosaurs in such numbers that the dinosaurs perished. A weakness of this theory was that crocodiles, which also lay eggs, survived. Another possibility was graphically depicted in the Disney movie *Fantasia*, in which the dinosaurs died of thirst under a scorching sun. A variant of this was the explosion of a nearby supernova, which showered the Earth with ultraviolet radiation that proved lethal to the dinosaurs. It was also known that massive volcanic eruptions occurred shortly before the extinction of the dinosaurs; perhaps this was in some way related.

In 1980, Walter Alvarez was examining a site in Italy, where the boundary between the Cretaceous and Tertiary periods (known as the K-T boundary) was clearly evident. Walter asked his father, Luis, to help him analyze a thin clay layer that marked the dividing line between the two periods. Luis Alvarez was well-positioned to perform this task, as he was a physicist who had won a Nobel Prize and was at the University of California at Berkeley, where he had access to the necessary equipment. With the help of Frank Asaro and Helen Michel, they discovered that the clay had a much greater concentration of iridium than normal, and this iridium concentration was to characterize the K-T boundary throughout the world.

The Alvarezes, Asaro, and Michel proposed that a large asteroid or comet, enriched in iridium, had collided with the Earth just at the end of the Cretaceous. The dust thrown up by this collision had stayed in the atmosphere, blocking the passage of sunlight and preventing plants from photosynthesizing, thus destroying the crucial first link in the food chain. With the cessation of plant growth, many species would be forced into extinction.

This attractive theory generated widespread appeal, but much work still had to be done to verify it. One vital link would be the discovery of a large impact crater at the appropriate time, and one such candidate has been found in the ocean off Mexico. Even though this theory has not yet been completely accepted—volcanic eruptions still have their supporters among paleontologists—it has forced a rethinking of the role played in the history of the Earth by catastrophes such as meteor collisions. These catastrophes may well have prompted many of the great mass extinctions that have occurred since the development of life on Earth.

This theory not only may explain the demise of the dinosaurs, but may, in some small way, have helped to prevent the extinction of man. After this theory was proposed, several scientists suggested that a nuclear war might well produce a similar effect, throwing enough dust and soot into the atmosphere to lower the global temperature significantly. This would have an extremely adverse impact on plant growth. This "nuclear winter" scenario was (and still is) the subject of serious investigation and debate, and the real possibility of such an occurrence may well have lowered the potential for a nuclear war. Now, however, it is the other side of the coin—the possibility of warmer weather produced by the greenhouse gases accompanying the burning of fossil fuels—that is of greater concern to the scientific community.

CHAPTER 7

Genetics and DNA

One of the great accomplishments of science has been to understand genetics—the mechanism of inheritance. This accomplishment—and what we do with it—may well be the standard by which our species will be measured. Understanding and controlling this mechanism enables us to make crops more productive, and to direct the future evolution not only of other species, but also ourselves.

The Laws of Genetics

Somewhere between ten and fifteen thousand years ago, mankind underwent perhaps the most significant change in history: changing from a society based on hunting animals and gathering edible plants to a society based on farming and domesticated animals. There is a tremendous advantage to be gained by having stronger and more productive domesticated animals. It was noticed early that larger and stronger animals tend to have larger and stronger offspring. However, the mechanism that created this was unknown at the beginning of the nineteenth century, at about the time that Johann Mendel, son of a Czechoslovakian farmer, was born.

Mendel's parents did their best to secure a good education for him by placing him in a monastery, where he took the name Gregor. The monks there sent him on to the University of Vienna to obtain a formal diploma as a teacher. Mendel, however, was a poor student, and his examiner, observing that he "lacks insight and the requisite clarity of knowledge," failed him.

At the time, it was generally believed that the mechanism governing inheritance worked by averaging: a red plant crossed with a white plant would have pink offspring, pink being the "average"' of red and white. On returning to the monastery, Mendel pursued a different viewpoint. He hypothesized that each inherited characteristic (now called "genes"), such as size or color, came in two varieties, which he called dominant and recessive. If an organism inherited a dominant gene from either parent, then that dominant gene would be expressed in the organism. Only if the organism inherited recessive genes from both parents would it display recessive characteristics.

To illustrate this principle, Mendel raised pure-bred tall and short peas, and then crossed them. The first-generation hybrids were all tall, each having inherited a dominant tall gene and a recessive short one. When the hybrids were crossed, roughly three-quarters of the plants were tall, and one-quarter were short; none were of intermediate height. Mendel constructed a mathematical model for this, pointing out that there were four possible second-generation plants. One would have inherited a tall gene from both parents, one would have inherited a tall gene from the male parent only, one would have inherited a tall gene from the female parent only, and one would have inherited a short gene from both parents. Since the tall gene was dominant, three-quarters of the second-generation plants should be tall.

These brilliantly constructed experiments, along with Mendel's explanations, were published in the *Journal of the Brno Natural History Society*, where they were immediately ignored by the scientific community. Mendel attempted to publicize them by sending copies of the article to several distinguished scientists. He might have continued his efforts along this line, but a rather unexpected development occurred: he was elected abbot of the monastery. Mendel took his duties seriously and mostly abandoned his research for the remainder of his life.

Then, in 1900, a remarkable coincidence occurred. Three investigators (Hugo de Vries, Karl Correns, and Erich von Seysenegg) independently rediscovered Mendel's results. They each instituted a thorough search of the literature, and all three discovered Mendel's prior work in the obscure journal in which it was published. In the best scientific tradition, when they published their results, they all gave credit to Mendel.

Mendel apparently did have some interest in pursuing his work on genetics, and after he was elected abbot of his monastery he tried to extend his method of experimentation with pea plants to the animal kingdom, choosing to work with bees. As a result, he developed a hybrid bee, which gave excellent honey. Unfortunately these bees were extremely ferocious, being much more prone to sting than the standard honeybee. The bees were subsequently destroyed. Mendel apparently was not only the father of genetics; he was the father of "killer bees" as well. One can only wonder if the Africanized "killer bees" that invaded southern California toward the end of the twentieth century would be doing so if Mendel's work in this area had been adequately publicized.

The Biochemistry of Genetics

Mendel's laws of genetics explained how characteristics were inherited. What was then needed was an explanation of the physical mechanism behind those laws. This physical mechanism was molecular, and it took almost a century before it was fully worked out.

CHROMOSOMAL INHERITANCE AND MUTATIONS

The middle of the nineteenth century saw two great revolutions in biology—Darwin's theory of evolution, and Mendel's laws of genetics—although the existence of the latter was only unearthed at the beginning of the twentieth century. There was still a great deal of debate concerning the validity of these theories, and much of it centered on the evolution of new species.

According to Darwin, natural selection was the driving force behind evolution. Natural selection might explain why antelope species became swifter, as the swifter antelopes would be the ones most likely to escape predators. Mendelian genetics could explain the proportion of blue-eyed children born to brown-eyed parents. However, neither theory seemed able to explain how an entirely new species, such as human beings, arose.

The Dutch biologist Hugo de Vries had shown that in one generation, large variations in plants could produce an entirely new species. De Vries felt that these large variations, which were called "mutations," could explain the existence of new species of animals. Thomas Hunt Morgan, an American geneticist who favored de Vries's theories, decided to test them.

To do so, Morgan worked with fruit flies. These insects had many advantages from the standpoint of genetics. They were small and they proliferated extremely rapidly, so many generations could be studied in a short period of time. Equally important, the cells of the fruit fly contained only four chromosomes. At the time that Morgan began his work, the chromosomes were suspected of carrying genetic information.

However, there was a problem with this theory. Humans have only two dozen chromosomes, yet there are thousands of inherited characteristics. If the chromosomes did indeed carry genetic information, there must be many such characteristics, called genes, on each chromosome. By using fruit flies and keeping careful track of the characteristics of each generation, and correlating this with the physical makeup of the chromosomes, Morgan was able to confirm that each chromosome did indeed carry a specific set of genes.

Morgan actually managed to do considerably more than confirm the chromosomal theory of inheritance. He was able to show that pairs of chromosomes occasionally switched portions of their material. This phenomenon, which Morgan called "crossing over," introduced a new variability into the biology of inheritance. Normal inheritance allowed for small changes from generation to generation; "crossing over" provided a mechanism for the significant changes de Vries had spotted in plants.

Through ingenious data handling and experimentation, Morgan was also able to pinpoint the physical location on the chromosome where a particular gene could be found. He was thus able to establish correlations between various genes. Today we know there are many such correlations in the human chromosomes. For instance, it is well known that men are much more liable to be color-blind than women. In 1911, Morgan began drawing up maps of the chromosomes of the fruit fly, becoming in the process the first gene mapper. Gene mapping is even more important and significant today in the era of genetic engineering; the successful Human Genome Project is merely the latest extension of Morgan's pioneering efforts.

Many individuals inherited not genes from Morgan, but his interest in biology. His student Hermann Muller managed to induce mutations in fruit flies by the use of X-rays, thus demonstrating yet another mechanism by which mutations could occur. Finally, his niece Isabel Morgan performed pioneering experiments in devising a vaccine to prevent polio in chimpanzees, and this work proved extremely important in Jonas Salk's development of a polio vaccine.

STRUCTURE AND SYNTHESIS OF AMINO ACIDS

By the middle of the nineteenth century, the investigation of the chemical structure of molecules associated with living things was well under way. The analytical techniques of the time could deduce the composition of molecules such as sugars and fats, but the proteins would turn out to be substantially larger, and consequently considerably more difficult to decipher.

Justus von Liebig, one of the great German chemists, had shown that the proteins were necessary to life, as they supplied substances such as nitrogen, sulfur, and phosphorus that are not present in fats or sugars. One chemical technique used attempted to break down a molecule by heating it in acid; this procedure decomposed cellulose into its glucose components. Henri Braconnot tried the same procedure on gelatine, obtaining a sweetish, crystalline acid he called "glycine." When he tried the same technique on muscle tissue he obtained another crystalline acid he called "leucine." Later analysis would determine that both these acids had a similar structure, including an amine group (NH_3). They were named "amino acids."

The list of amino acids grew. Liebig would also discover tyrosine. Threonine, the last of nineteen nutritionally important amino acids, was discovered by William Rose in 1935. By the end of the nineteenth century, the chemists were certain that the building blocks of proteins were amino acids, just as the building block of cellulose had been shown to be glucose. However, there was only one glucose molecule, but there were many different amino acids.

All the amino acids had two features in common—the amine group, and a carboxyl group. Emil Fischer, a brilliant German chemist, believed that proteins were constructed by stringing together amino acids, the amine group at the head of one amino acid joining the carboxyl group at the tail of the next by means of a bridge he called a "peptide bond." By 1907, Fischer was able to synthesize a sequence of eighteen amino acids linked by peptide bonds. However, this was not a true protein in the sense that it did not exist in nature. Twenty-five years later Fischer's student, Max Bergmann, conclusively demonstrated the truth of Fischer's theory. Bergmann showed that natural digestive enzymes, which broke down proteins into their component amino acids, attacked Fischer's synthetic molecules in exactly the same manner. The digestive enzymes are specifically tailored to cut only one type of bond, so the bond joining the amino acids in the synthetic molecules must have been the same one that joined the amino acids in the proteins.

William Rose, who had discovered the last of the nutritionally important amino acids, devoted his career to demonstrating which amino acids were necessary for the diet, and precisely what quantities were needed. He also discovered which amino acids the body could synthesize, and which must be supplied through diet. Rose's work was a classic example of science in the public interest. As a result of Rose's investigations, it is known that certain amino acids are not present in vegetables, and anyone who follows a vegetarian regime must be sure to supplement the diet with the missing amino acids.

Emil Fischer's work was also one of the foundations of the modern pharmaceutical industry. He synthesized barbituric acid, which is used as a sedative, and many commonly used sedatives are barbiturates. Fischer, whose professional career was marked by a string of successes, was to encounter great tragedy in his personal life. He organized food and chemical production for the German government during World War I, during which two of his three sons were killed. Depressed by this, as well as the fact that he was suffering from cancer, he committed suicide in 1919.

THE STRUCTURE OF DNA

The search for the mechanism behind inheritance resembled the plot of many of the classic movies of the last quarter of the twentieth century. It was *Raiders of the Lost Ark*, with the prize the secret of life itself. It was *The Odd Couple*, with as unlikely a pair of protagonists as the ones Neil Simon constructed. It was *Rocky*, with the challenger going up against the reigning champ. Throw in a little sex and a brilliant and talented woman doomed to a tragically early death, and you have the elements of one of the great dramas—scientific or otherwise—of all time.

The story starts, as previously discussed, with Gregor Mendel, the Austrian monk who discovered the laws governing inherited characteristics. Soon after, scientists began the long struggle to discover the mechanism by which Mendel's laws were enacted. In 1944, Oswald Avery, Colin MacLeod, and Maclyn McCarty demonstrated that deoxyribonucleic acid (soon to be universally known as DNA) was the substance through which inherited characteristics were transmitted.

This was only the first piece of the puzzle. Still to be determined was precisely *how* DNA did what it did. By 1951, the chemical composition of DNA had been ascertained—it consisted of sugars and phosphates, which are fairly simple compounds, and four bases: adenine, cytosine, guanine, and thymine. It was also known that, even though different samples of DNA might have differing amounts of the four bases, the amount of adenine was always the same as the amount of thymine, and the amount of cytosine was always the same as the amount of guanine.

Hot on the trail of the structure of the DNA molecule was Linus Pauling, unquestionably the world's leading biochemist. Almost complete newcomers to the problem were James Watson, a young American postdoctoral student recently arrived in England, and Francis Crick, a somewhat older, British graduate student with a background in mathematics. Despite the fact that they knew Pauling was undoubtedly the frontrunner, Crick and Watson decided to enter the race to decipher the structure of DNA.

They had two unusual allies: Maurice Wilkins and Rosalind Franklin. These two were trying to work out the structure of DNA by using X-ray diffraction photographs of DNA crystals, a technique consisting of examining how X-rays bounced off DNA. While these four were trying to put the pieces together, Pauling wrote an article in which he claimed to have worked out the structure of DNA as a triple helix. When Watson and Crick reconstructed Pauling's model and showed it to Franklin, she pointed out that it disagreed with her diffraction data. Since Pauling was backing triple helixes, Watson decided that his best bet to beat Pauling would be to experiment with double-helix models.

One day, in a flash of insight, Watson realized that the shape of an adenine-thymine pair would be the same as the shape of a cytosine-guanine pair. That would account for the equality between the amounts of adenine and thymine, and cytosine and guanine. After several false steps, Watson came up with a double helix model incorporating these features, and Crick made the calculations to demonstrate the feasibility of the model. Wilkins and Franklin made X-ray diffraction computations that substantiated the model. Proteins do most of the work in an organism, and the secret of life—how the cells know which proteins to produce—had finally been discovered.

After learning of the double-helix model, Pauling visited Cambridge in the spring of 1953. He acknowledged the error in his thinking that had led to his construction of an erroneous model, and agreed that the double-helix model of DNA was undoubtedly correct. The very next year, Linus Pauling would win the Nobel Prize in Chemistry for previous work done on chemical bonds. Crick, Watson, and Wilkins would share the Nobel Prize in Physiology or Medicine in 1962. Franklin, whom all agreed had made substantial contributions, had tragically passed away from cancer at the age of thirty-eight, and was thus ineligible to share in an award that she richly deserved.

DECIPHERING THE GENETIC CODE

When Watson and Crick constructed the double-helix model for DNA in 1953, they also noted that the proposed structure would account for the process of cell duplication. Each strand of the double helix would unwind, and each of the approximately 4 billion bases on a strand would seek its complementary partner (adenines with thymines, cytosines with guanines) to reconstruct the other strand of the helix. Then, as a cell split into two, each strand would be able to generate a new molecule of DNA for each of the daughter cells.

The mechanism of heredity was now known. What was not known was how the molecule of DNA directed the essential process of life: the manufacturing of proteins. The genes that Mendel had described a century earlier were now known to have precise locations on a molecule of DNA, and each gene had a specific function: to manufacture a particular protein. Somehow, the sequence of bases in DNA directed the manufacture of proteins.

The obvious idea was that the 4-billion-long sequence of bases was a kind of computer program directing how proteins would be manufactured. Proteins were manufactured in a portion of the cell called a ribosome. Proteins themselves are chains of simpler chemicals called amino acids; every protein, from the hemoglobin in our blood to the pigments in our eyes, is constructed in an assembly-line fashion from a fundamental stock of twenty different amino acids.

The construction process actually involves more than just DNA itself. When the two strands of the DNA helix separate, a strand may be used as a template for the construction of another DNA molecule, but it may also be used to construct a molecule known as messenger RNA (abbreviated mRNA). The mRNA molecule is what is actually read by the ribosome, and differs from the DNA molecule in that the base uracil is used in place of thymine. As a strand of mRNA is fed to the ribosome, the ribosome sees a sequence of approximately 4 billion bases, a section of which might look like . . . UCGAGGUUCA . . . How does the ribosome know what to do with this sequence of letters?

The simplest idea would be that the ribosome sees a group of letters as a word, and each word as an instruction to attach a specific amino acid to the growing protein chain. There were only sixteen (4 × 4) different possible two-letter words made from the letters A, C, G, and U (AA, AC, AG, AU, . . . UA, UC, UG, UU). Since there were twenty different amino acids used in proteins, there would not be sufficient words in the instruction set to specify all the amino acids. There were sixty-four (4 × 4 × 4) different three-letter words, which would later be known as codons, that could be made from the letters A, C, G, and U, so the next step was to see whether a particular codon caused a specific amino acid to be added to the protein chain.

Marshall Nirenberg, a biochemist at the National Institutes of Health, was one of many scientists working on this hypothesis. In 1962, he managed to construct a section of mRNA consisting only of uracil bases. When this section of mRNA was read by a ribosome, it added the amino acid phenylalanine to the protein chain. The first codon in the genetic code had been deciphered: UUU stood for phenylalanine. Nirenberg would win the Nobel Prize in 1968, and within a few years, the genetic code had been completely deciphered.

One of the beautiful aspects of science is that information uncovered in one area can easily have ramifications in other areas. There is no a priori reason why each species cannot have its own genetic code: UUU could in theory code for phenylalanine in aardvarks, leucine in roses, and glycine in zebras. In reality, it does not work out that way: UUU codes for phenylalanine in all forms of life. This is powerful evidence for the theory of evolution, as the obvious way for this to occur is that the genetic code originally evolved in the simplest life-form and was passed from organism to organism even as the other genetic changes constituting evolution were taking place.

THE FUNCTIONING OF RNA

As might be suspected from the acronyms, DNA and RNA are closely related. The structure of DNA consists of a double-helix backbone to which are affixed four bases: adenine, cytosine, guanine, and thymine. The structure of RNA is quite similar, but it uses uracil instead of thymine. The chief functional difference is that, although DNA may get all the accolades, when it comes to the actual manufacture of the proteins coded for by the DNA, RNA does most of the work.

The existence of RNA has been known since the early portion of the twentieth century, and its importance in genetics long suspected. Indeed, even as James Watson and Francis Crick were on the verge of deciphering the structure of the DNA molecule, they were speculating on the role of RNA in the manufacture of proteins. They conceived of a sequence of operations that has come to be known as the central dogma of molecular biology. The DNA in the nucleus of the cell would act as a template for the manufacture of RNA. The RNA would then go from the nucleus of the cell to the cytoplasm, where it would direct the manufacture of proteins. This picture of the roles of DNA and RNA has been for the most part confirmed.

However, RNA has been discovered to play many other roles in the drama of life. The simplistic picture that DNA is copied to RNA, which is mechanically reproduced to create proteins, has had to be altered. The first major change in this picture was discovered by the French biochemists François Jacob and Jacques Monod, who discovered that chemical signals inside the cell determine whether or not the instructions within a gene are copied into RNA. This showed that the copying process from DNA to RNA involves editing as well.

A major part of editing, both in a manuscript and in a genetic molecule, is the removal of unusable material. The unusable material, called "introns," are cut out of the RNA message. The spliced segments must be rejoined, and this is accomplished by another form of RNA; these molecules are called "spliceosomes."

It still remains for the proteins to be constructed. This is accomplished by yet another form of RNA, called transfer RNA, whose job it is to read the final edited RNA message, fetch the required amino acids from the cytoplasm, and string them together to form the appropriate protein.

The threat of AIDS (acquired immunodeficiency syndrome) that emerged during the last two decades of the twentieth century brought about intensive study of the human immunodeficiency virus (HIV) that had been shown to be the cause of the disease. HIV is one of a class of viruses called retroviruses. A retrovirus is basically a strand of RNA encased in a coat of protein. Like other viruses, it cannot reproduce on its own, and must commandeer the genetic machinery of a cell to perform this task. In 1970, the American molecular biologist Howard Temin showed that retroviruses reproduce in a host cell by using the enzyme reverse transcriptase that is present in a cell to copy the RNA to DNA. The cell then copies the DNA into the RNA genetic material for the retrovirus, and also uses the DNA to produce the protein coat in which the strand of RNA is encased.

Which came first, the chicken or the egg? In molecular biology, the question is, which came first, DNA or RNA? Because RNA basically reads DNA, the prevailing impression early on was that RNA-based life was impossible. However, Thomas Cech showed that RNA could act as a catalyst and initiate changes on itself. This brought about the possibility of an RNA world, in which a simpler RNA-based life existed prior to DNA. In this scenario, DNA evolved later, and the more complex central dogma enabled more complex life to evolve.

GENETIC ENGINEERING

The attempt to alter existing life-forms for the good of man has been going on almost as long as recorded history. Breeding horses, dogs, and cattle date from well before the birth of Christ, and the attempt to produce new varieties of plants or ones that will grow better probably started at the same time that agriculture was developed.

In a sense, all of the above qualify as genetic engineering, but it is genetic engineering on a somewhat haphazard level, relying on the mechanism of chance to produce an improvement, and choice (known as artificial selection) to use this improvement. When Watson and Crick discovered the structure of DNA, the genetic material whose instructions are followed by the cell to produce proteins, it was immediately realized that the ability to manipulate DNA would greatly increase the power to improve existing life-forms or create new ones.

In 1968, Werner Arber, a Swiss microbiologist, was investigating a family of viruses called bacteriophages. These viruses actually eat bacteria (*phage* is Greek for "eat"). In the eternal struggle that characterizes natural selection, some bacteria are able to defend themselves against bacteriophages by producing a substance that prevents the growth of the viruses. Arber was able to show that the substance was an enzyme that actually cut the DNA of the bacteriophage at a specific location. Arber realized that this substance, which he called a restriction enzyme, located a specific sequence of molecules in the DNA strand, and would work nowhere else. These restriction enzymes were tools for cutting DNA at a precise spot, and are the workhorses of genetic engineering.

Five years later, Herbert Boyer and Stanley Cohen of the University of California at San Francisco performed the basic experiment that launched the genetic engineering revolution. Working with the common bacteria *E. coli*, they isolated a plasmid (a circular loop of DNA) containing genes enabling the bacteria to resist certain antibiotics. They used restriction enzymes to cut the DNA, and then inserted sections of other plasmids. This created a new plasmid, which contained genes from both of the original plasmids. When this plasmid was reinserted into an *E. coli* bacterium, the bacterium could be shown to display the genetic characteristics of both plasmids from which the engineered plasmid had been produced.

We are a half-century into the science of genetic engineering, yet it is already a multibillion-dollar industry with the promise to transform the world. The gene is the unit of inheritance. Genetic engineering holds the promise of producing plants that can manufacture their own fertilizer, flower more frequently, and ripen faster. On the human level, genetic engineering may make it possible to prevent a large number of so-called genetic diseases, which are caused by the inability of a gene to perform a desired function.

Of all the developments in this book, this is the one that is most likely to have a major impact on the ultimate development of the human race. Other developments will make it possible for us to change the Universe, but genetic engineering could make it possible for us to alter the evolution of humanity.

In her novel *Frankenstein*, Mary Shelley gave the world the caricature of the mad scientist, out to experiment with forces beyond his control. In general, scientists are much more conscious of the consequences of their actions than Shelley had suggested. Immediately after the discovery of genetic engineering, the biochemists met and instituted an extremely stringent set of regulations governing experiments in this area. The fear that a mutant strain of genetically engineered bacteria would wreak untold damage still exists, but for the most part the products of genetic engineering have proved more fragile than their naturally existing counterparts. As a result, the initial stringent regulations governing genetic engineering experiments have been substantially eased. GMO—genetically modified organisms—are now a staple of everyday life.

The Big Question: the Origin of Life

If there is one question whose answer would interest both scientist and non-scientist alike, it is this: How did life begin?

Until the nineteenth century, this question was not even in the scientific domain, as it was assumed that life was created by a supreme being. With the demise of vitalism, and the obvious age of the many fossils that were being discovered, the origin of life began to be the subject of serious scientific inquiry.

It is still an unanswered question, but like much of science, it is a work in progress. It involves exploration, experimentation, and theorizing.

Until the middle of the twentieth century, the oldest life-forms that had been discovered were fossils whose age was estimated at approximately 500 million years. In 1954, Elso Barghoorn headed an expedition to study the Gunflint chert, an ancient rock formation in Canada. When portions of the rock were sliced and examined under a microscope, they revealed fossils similar to modern bacteria and blue-green algae. This discovery pushed back the earliest known appearance of life to approximately 2 billion years ago. Almost fifteen years later, Barghoorn discovered amino acids in rock whose age was 3 billion years. The limit has recently been pushed back even further with the discovery of traces of bacteria appearing to be approximately 3.9 billion years old.

This is both encouraging and discouraging. The surface of the Earth is believed to have been molten approximately 4.2 billion years ago, which gives life only 300 million years to get going from a standing start. While it is encouraging that life seems to have started fairly easily, it is discouraging because organic molecules such as DNA are extremely complex, and it seems highly unlikely that they could form by chance.

In 1953, Stanley Miller was a graduate student in the chemistry laboratory of Harold Urey, a Nobel Prize–winning chemist. Urey suggested that Miller simulate the atmosphere of a prebiotic Earth in which no oxygen existed, so Miller concocted an atmosphere consisting of chemicals presumed to exist on the early Earth: water, methane, ammonia, and hydrogen. He also assumed that there would have been plenty of lightning, so he supplied an electrical spark to mimic lightning. He sealed the flask, waited a week, and observed an orange gunk on the surface of the flask. The gunk proved to contain amino acids, which are critical to life as we know it.

Miller's experiment showed that complex organic molecules could arise from relatively simple ones. Of course, amino acids are not alive, but a week is not 300 million years, either. Many Miller-like simulations have been done, with various atmospheres and varying conditions, but no one has yet opened a flask and found a living entity.

There are numerous theories on how life could have arisen. The Russian biochemist Alexander Oparin is the author of the "warm little pond" theory, in which somehow enough organic molecules got trapped in a suitable environment for life to arise. Some theorists feel that underseas hydrothermal vents, the "black smokers" described in the previous chapter, may have provided a suitable environment. There are those who feel that the evolution of life was a two-step process, assisted by some transitional inorganic form. The British crystallographer John Desmond Bernal favored mineral catalysis, while A. Graham Cairns-Smith of the University of Glasgow believed that inorganic clays could serve as a scaffold on which organic molecules could be built—after the molecules were capable of self-reproduction, the scaffolding disappeared.

One thing is certain: if the answer is discovered and proven to be correct, it will be one of the most interesting scientific achievements of all time.

An indication of how serious is the search for the answer to this question can be found in the fact that NASA has funded interdisciplinary investigations on this subject. Whether there is life on other worlds is still an unanswered question, and NASA's thinking is that we will be more able to recognize life on other planets if we know how it arose on this one. It's been nearly three-quarters of a century since the Miller–Urey experiment, and we still don't have an answer. We have to entertain the possibility that, although we may be able to devise a scenario by which life could begin, we may never know how it actually did begin.

CHAPTER 8

The Human Body

Our curiosity about the nature of our bodies begins almost immediately after we are born, and continues throughout our lives. Our bodies are astounding in so many ways—the conscious ways we can direct them, the systems that billions of years of evolution make function so well—systems of which we are largely unaware. What is also amazing is the range of human capability, such as the abilities of a concert pianist or a topflight athlete. This range also includes the incredible accomplishments of the human brain—some of which are in this book.

Our bodies do many things well—but, as many have observed, there is no physical ability that humans possess that is not exceeded in some other species. We're generally faster than turtles, and slower than cheetahs. But no other species has the brain that enables us to surmount our inabilities—and enrich our lives, a concept unknown to any other species.

Structure and Organs: What the Eye Can See

The science of the human body began with what was accessible to the human eye—itself one of the most amazing products of evolution. This was essentially the only tool available by which data about the human body could be acquired. Knowledge of the structure and function of the various portions of the human body was valued in many cultures for the obvious reason that this knowledge could help to repair damage—but the acquisition of that knowledge sometimes ran into societal and religious roadblocks. Such was the case during the Dark Ages—a thousand-year period during which investigation into the structure and function of the human body advanced only

147

infinitesimally, at least in Europe, although progress was made in this area in other societies. Nonetheless, when the prohibitions—both actual and tacit—were lifted, the value of acquiring such knowledge, long evident, was encouraged.

ANATOMY

Considering how little was actually known about the body and the nature of disease in ancient Greece, it is amazing how rational some of the ancient physicians appear by modern standards. It is surprising to learn that the sayings, "One man's meat is another man's poison," and "Desperate diseases require desperate remedies," are actually attributed to Hippocrates, the father of medicine.

Hippocrates was in fact better known to the ancients because he founded a school of medicine, rather than because he was a doctor. Perhaps his greatest contribution to medicine was not the Hippocratic Oath, but rather the view that disease was a physical phenomenon, rather than the result of having incurred the wrath of some deity. However, the next great physician would not appear until several hundred years after Hippocrates's death. That physician was Galen, who climbed the ladder of professional success until he became court physician to Emperor Marcus Aurelius.

Galen was the greatest anatomist in the ancient world, but he had the misfortune to live in an era in which human dissection was no longer being practiced. As a result, Galen would dissect animals, observe what he could, and generalize to human beings. He was the first to identify many of the major muscles, and also showed the role of the spinal cord by severing it in animals and noting the ensuing paralysis.

Galen was certainly the most influential physician of his time. His extensive writings were carefully preserved throughout the Dark Ages, partly because his religious views were in line with the Christian thought that everything in the Universe was designed for a purpose. When the search for new knowledge ceased shortly after the fall of Rome, Galen represented the state of medical art, and remained the unquestioned authority in the field for more than a thousand years.

By the sixteenth century, though, there was a new willingness to doubt some of the previously unquestioned authorities. One such doubter was Andreas Vesalius. Though raised in France, he had relocated to Italy, where there was a greater spirit of intellectual freedom. One consequence was that the practice of human dissection, though nominally forbidden, was now standard at Italian medical schools.

Vesalius eventually obtained teaching positions at many of the leading Italian universities. However, the prevailing practice was for the teacher to lecture while the assistants did the dissections. Vesalius was disgusted with the poor job performed by his assistants, and took over the job of doing both dissections and commentary himself.

Vesalius was a man of extraordinary skill as a lecturer and teacher, but his greatest contribution was a single book that revolutionized anatomy. This book, *De Humani Corporis Fabrica* (On the Structure of the Human Body), was not only the first great work on anatomy, but contained accurate illustrations and was printed, rather than being handwritten. This enabled many copies of the book to be produced, and printing assured that the reproductions were accurate, which was not always the case when a book was copied by hand.

As a result of this book, not only were many of Galen's errors corrected, but the reliance on Galen's theories and practices ended as well. Before Vesalius, medicine had been static for more than a thousand years. Vesalius's book, and the changes it brought about, made it possible for medicine to advance.

By the time *De Humani Corporis Fabrica* was published, mechanized printing had existed for almost a century, yet many books were still just printed versions of handwritten manuscripts that had been around for centuries. Vesalius's book was published in 1543, which may have been the first truly landmark year for the publishing business. In the same year, Copernicus also published his book, in which he presented the heliocentric theory of the solar system.

THE CIRCULATORY SYSTEM

The crucial role of the heart has been intuitively understood since the earliest days of recorded history. In ancient Egypt, the fate of the soul was thought to be determined by the weight of the heart. Egyptian priests would weigh the heart of the dead on a scale against a feather, believing that those who had hearts that were not "heavy with sin" went on to happiness in the afterlife.

Aristotle, one of the greatest intellectual giants in history, thought that the heart was the seat of the soul, and attributed mystical powers to it. In 130 CE, Galen, personal physician to the Roman Emperor, advanced the concept of a circulatory system through which blood flowed from the body to the heart and back to the body. However, medical research did not occupy an important place in Roman society, and when Rome collapsed and the Dark Ages began, medical research was basically "put on hold" throughout Europe.

As Europe emerged from the Dark Ages, people began displaying a greater interest in medicine, spurred on by the carnage wrought by the Black Death. Once again, though, interest in the functions of the human body ran afoul of the proscriptions of the Church. Human anatomical investigations and drawings were strictly forbidden—indeed, Leonardo da Vinci had been forced to steal corpses in order to make his accurate anatomical drawings. The Italian anatomist Vesalius had to have his groundbreaking book on human anatomy printed in Switzerland in order to avoid being condemned, excommunicated, or worse by the Italian authorities. This fear was well-founded. When Miguel Servetus published a book containing conjectures on the role of the heart in pumping blood, he was burned at the stake by the Spanish Inquisition, with a copy of his book tied to his body.

At the start of the seventeenth century, William Harvey, a well-to-do Englishman who had studied at Cambridge, went to Padua to study medicine, as Padua had for three centuries housed the finest medical school in Europe. While Harvey was in Padua, Galileo's experiments in mechanics and astronomy were setting a new standard for science. On his return to England, Harvey decided to apply Galileo's methods to the study of the heart and the circulation of the blood.

His principal tool was dissection; in his attempts to understand the heart, he is said to have dissected over eighty species of animals. Harvey determined that the heart is a muscle, and that it operated by contraction. He calculated the rate at which the heart pumped blood, and determined that in one hour the heart pumped a quantity of blood that was about three times the weight of a human being. Since it seemed impossible to construct a mechanism that would destroy and recreate blood at such a rate, the obvious conclusion was that the blood was being circulated throughout the body.

Harvey further noted that the valves in the arteries and the veins were one-way valves, and then observed that blood flowed away from the heart through the arteries, and toward the heart through the veins. Although his results initially met with substantial opposition from the medical establishment, Harvey himself refused to debate the matter, publishing a book on the subject and letting the facts speak for themselves. Within a generation, his results were universally accepted, and the groundwork for the study of physiology had been established.

There was an obvious problem in Harvey's theory of blood circulation: how did blood, which flowed *from* the heart through the arteries and *to* the heart through the veins, make the jump from the arteries to the veins? Harvey, noting that both arteries and veins subdivided into blood vessels with smaller and smaller diameters, applied inductive reasoning, deducing that the actual connections were too fine to be seen. Four years after he died, the Italian physiologist Marcello Malpighi observed the connecting blood vessels with the aid of a microscope.

What the Eye Doesn't See

It's hard to imagine where science would be without the microscope. This amazing tool not only revealed unseen worlds, but greatly aided our understanding of ourselves.

Life has had almost 4 billion years to evolve on Earth, and given that much time, it isn't surprising that it has come up with systems of amazing complexity and functionality. Even though we think of the human brain as the apex of the evolutionary pyramid, many of the systems that enable us to live our lives take place without conscious direction. And a good thing, too, because we are probably incapable of consciously directing even such a simple process as the digestion of food, to say nothing of something so complex as the functioning of the immune system.

NERVES AND NEUROTRANSMITTERS

The Greeks are widely recognized as having produced excellent playwrights, philosophers, and geometers, but it comes as a surprise to realize that their knowledge of anatomy was rather sophisticated. The anatomist Herophilus, who lived three centuries before Christ, was quite interested in the brain and nervous system. He classified nerves into two categories: the sensory nerves, which received impressions from the five senses, and the motor nerves, which stimulated motion. For more than two thousand years, this would represent the most advanced thinking on the subject.

The Romans were not much interested in theoretical science of any sort, and the Dark Ages produced almost no scientific advances. The Renaissance brought about a reawakening of interest in natural phenomena of all sorts. In 1771, Luigi Galvani made one of the most important observations in the history of science. He noticed that some dissected frog legs twitched excitedly when struck by an electric spark. In a sense, this was not altogether surprising, because live muscles were known to twitch when subjected to electricity. Nonetheless, this discovery was to inaugurate many different lines of research, one of which we have already seen in chapter 5.

One of these lines of research, which had been dormant for two thousand years, was on the interaction of nerves and muscles, and the nature of the nervous impulse. In 1826 Johannes Müller, a German biologist, was experimenting with sensory nerves. It was well-known by that time that light stimulated the optic nerve, and that this stimulus was interpreted by the brain as visual brightness. Müller discovered that if the optic nerve was stimulated by electricity, that stimulus would still be interpreted as visual brightness. He later showed that this was true for any type of nerve; no matter how it was stimulated, the brain would always interpret it the same way. Not only did this indicate that nerves functioned through electrical impulses, it greatly simplified the investigation of the nervous system.

A century later, it would be discovered that the nervous system was not simply an electrical circuit. Otto Loewi, a German physiologist, was studying the nerves of a frog's heart. He had discovered that certain chemical substances were released when the nerve was stimulated. One morning he awoke at 3:00 a.m. with the idea for a brilliant experiment, which he then wrote down. The next morning he couldn't read his own handwriting! That evening, he again woke up at 3:00 a.m., and corrected his previous error by immediately going to the laboratory and conducting the experiment, which was to take the chemical substances that had been released and show that these had the power to stimulate heart muscle without the intervention of a nerve signal.

During the previous decade, the British biologist Henry Dale had been working on fungi, and had isolated a compound called acetylcholine. This substance had an effect on organs similar to the effects produced by reception of nerve impulses. When Dale read of Loewi's experiment, he was able to show that the substance Loewi had discovered was acetylcholine. Acetylcholine was the first neurotransmitter, a class of chemical compounds that inhibit and excite the transmission of nervous impulses.

Neurotransmitters have been shown to be important components of behavior. In 1972, a team of medical researchers discovered that bipolar disorder (often called "manic depression") is the result of an imbalance between two types of neurotransmitters. As a result of this discovery, it has been possible to treat several types of behavioral disorders by chemical means.

For their work on acetylcholine, Loewi and Dale received the Nobel Prize, which probably saved Loewi's life! Loewi was Jewish, and when Hitler invaded Austria, he was arrested. However, the Nazis may have realized that it would have been bad public relations to execute an eminent scientist, and Loewi was allowed to leave the country provided he turn over his share of the Nobel Prize money to the Nazis.

THE FUNCTIONING OF THE IMMUNE SYSTEM

The immune system is one of the great mechanisms of survival. Even before AIDS and COVID-19, the importance of the immune system was clear. The immune system is the body's intricately organized defense against foreign invasion, and learning how the immune system works, how to strengthen it, and what its weaknesses are has been and will be critical to the advancement of medicine.

The basic mechanism of the immune system is that, once it is exposed to a foreign substance, it learns how to manufacture defenses against it. Edward Jenner unwittingly exploited this mechanism when he inoculated people against smallpox by giving them a mild case of cowpox. Once the germ theory of disease became accepted, progress in the understanding of the immune system accelerated.

One of the first great developments in understanding immunity was the result of an experiment in 1890 by Emil von Behring and Shibasaburo Kitasato. They injected guinea pigs with blood from other guinea pigs known to be immune to diphtheria, and observed that the injected animals acquired that immunity. As a result, they concluded that immunity was conferred by protective substances in the blood, which von Behring called antibodies.

Paul Ehrlich, the originator of chemotherapy, was inclined to chemical explanations for all biochemical phenomena. He suggested that an antigen, a substance that provoked a reaction from the immune system, had a specific molecular structure, and that the antibody manufactured by the immune system fit the antigen much like a key fits a lock. This insightful theory was later confirmed by Karl Landsteiner, who showed that in order to combat a specific antigen, the immune system manufactures a specific antibody. Landsteiner not only confirmed Ehrlich's theory, but he also used the antigen-antibody reaction to develop the system of blood typing that we use today. Landsteiner's blood typing makes it possible to give blood transfusions without provoking an undesirable response from the immune system.

The antibody–antigen reaction is not always beneficial. Allergic reactions occur when the immune system produces antibodies to substances that are not intrinsically harmful. Skin grafts and organ transplants are frequently met with a reaction known as rejection; the body tries to destroy the graft or the transplant. Peter Medawar showed that rejection was an immune system response to the new material. In studying this phenomenon, Frank Burnet observed that a developing organism produced antibodies only in response to antigens that it encountered later in its life, and suggested that the immune system ignores antigens it encounters early in life. Although Burnet was unable to prove this, Medawar was able to do so.

Burnet eventually devised the clonal selection theory to explain immune response. When a lymphocyte initially encounters an antigen, it multiplies and produces identical lymphocytes, which manufacture the antibody necessary to counteract that particular antigen. Later research has proved that Paul Ehrlich, possessed of one of the keenest chemical intuitions, was correct: an antigen recognizes an antibody by identifying specific patterns on the surface of antigen molecules.

The 1980s saw the onset of the AIDS epidemic. AIDS is an insidious disease caused by HIV (human immunodeficiency virus). Viral diseases are especially hard to combat because they are often impossible for the immune system to detect. A virus is simply a strand of genetic material enclosed in a coat of protein. The coat itself is innocuous; but when the virus gets inside a cell, the genetic material of the virus can commandeer the cell's own genetic machinery and reproduce the virus. Worse, the HIV kills cells. The year 2019 brought with it the COVID-19 pandemic, also a viral disease, but fortunately one for which it was possible to develop a vaccine. HIV has many more evasive strategies available to it than the COVID-19 coronavirus, but as of this writing there is an ongoing trial for an AIDS vaccine based on the same methodology as the successful mRNA vaccines used against COVID-19.

Biochemistry of the Human Body

There has been a lot of cross-pollination in the sciences over the past fifty years. Stephen Hawking, probably the best-known scientist of the last half-century, was an astrophysicist—a discipline that didn't exist until the second half of the twentieth century. But biochemistry has existed for substantially longer. As scientists came to realize, biology at its core is chemistry, and many of the great advances in our understanding of how the human body functions are essentially comprehending the structure and functions of the highly complex molecules that are necessary for life.

THE STRUCTURE AND FUNCTION OF HEMOGLOBIN

The importance of blood to life has been known for thousands of years. To the Greeks, who formulated the first theory of living organisms, blood was one of the four humours, along with phlegm, black bile, and yellow bile. It was obvious to even a child that blood was important, but for thousands of years, no one knew what function blood performed in an organism.

It was known that blood was not a simple liquid. A container of blood, left unattended, would separate into a red liquid and a pale yellowish fluid, separated by a thin layer of white. The first great breakthrough in discovering the nature of blood could not be made until the seventeenth century, after the microscope had been invented. Although the Dutch microscopist Jan Swammerdam had observed red blood cells in frogs as early as 1658, he did not publish his results. Fortunately for science, Anton von Leeuwenhoek had observed red blood cells in human blood, and described his results in 1673.

The role that red blood cells played in the body was not discovered for almost two centuries, when the idea of counting blood cells as a measure of health was devised by François Magendie. This was a period right after the vitalistic view of organisms had been overturned, and researchers were quantifying many aspects of the diagnostic procedure.

Along with improved diagnosis came improved analytical methods. One of the chief new analytical tools of the era was the spectrometer, an instrument whose impact on both chemistry and physics has been profound. While astronomers were attaching spectrographs to telescopes to decipher the composition of distant stars, biochemists were using them to analyze many of the substances that are found in organisms.

One of the leaders in this area was a German biochemist, Ernst Hoppe-Seyler. It was Hoppe-Seyler who gave the name "hemoglobin" to red blood cells, and who demonstrated that the function of red blood cells was to transport oxygen. Hoppe-Seyler crystallized hemoglobin (many proteins can be crystallized), and also was the first to notice the sinister fact that carbon monoxide could readily be transported by hemoglobin. He was also the first to notice chemical similarities between hemoglobin and chlorophyll.

Hemoglobin is an extremely complicated molecule. Its exact structure could not be determined until 1960, when Max Perutz used X-ray crystallography, high-speed computers, and the ingenious trick of adding a single heavy atom of gold or mercury to the molecule to help clarify its structure. Anyone who observes the molecular structure of hemoglobin and chlorophyll cannot help but come to the conclusion that, because their forms are so similar, their functions must be as well.

In more than one sense, Hoppe-Seyler could be called the father of biochemistry. In addition to his own work, he established the first journal of biochemistry, and some of his students were pioneers in the field. The most famous was probably Johannes Miescher, who was the first to discover the nucleic acids (acids appearing in the nuclei of cells), of which DNA and RNA are undoubtedly the most well-known examples.

There are actually three types of blood cells, and each has an important role to play. The red blood cells transport oxygen. Élie Metchnikoff discovered that white blood cells immediately move to damaged areas of the body, and also attack and devour invading bacteria. The third type of cell, the platelets (not surprisingly, these are shaped like little plates), were found by Giulio Bizzozero to play a key role in blood clotting. All of the body's blood cells are created from parent cells in the bone marrow.

THE DISCOVERY OF HORMONES

Ivan Pavlov is one of the more well-known names in science. It is somewhat ironic that he won a Nobel Prize for research with which very few people are familiar, but nonetheless was indispensable to the work for which he is best known.

Pavlov was the son of a priest, and initially studied to become a priest. At the theological seminary, he read Darwin's *Origin of Species*, and decided to switch to science.

If you were a dog in Russia, you probably wanted to stay as far away from Pavlov as possible, as most of his experiments used dogs as laboratory animals. His initial work involved showing that the stomach's gastric juices were stimulated not by the arrival of food in the stomach, but by signals sent from the brain. This research was crucial in establishing the importance of the autonomic nervous system, and won Pavlov the Nobel Prize in physiology and medicine.

The experiment for which he is best remembered was actually an offshoot of his previous work, which showed that the stimulation of nerves in the mouth by food brought about a response in the stomach. This mechanism is known as an unconditioned reflex, a class of behaviors with which organisms are born.

Pavlov knew that the salivation of a hungry dog when shown food was an unconditioned reflex. He decided to see whether it would be possible to induce a reflex that was not present at birth. To do so, he rang a bell every time the dog was shown food. Eventually, the dog would not only salivate in response to the food, but in response to the bell. This is known as a conditioned reflex, and conditioned reflexes would play a key role in the behaviorist theories of psychology.

Pavlov's earlier studies came to the attention of William Bayliss and Ernest Starling, two English physiologists. Pavlov's earlier results had led Pavlov to believe that many digestive reactions were controlled by the nervous system. Bayliss and Starling studied how the pancreas began to secrete its digestive juice when acidic food contents passed from the stomach to the intestine.

Bayliss and Starling tried to confirm Pavlov's hypothesis by cutting the nerves to the pancreas. To their surprise, the pancreas continued to secrete its digestive juice. Further investigation revealed that the lining of the small intestine secreted a substance, which they called secretin, when it was exposed to stomach acid. It was secretin that stimulated the pancreas to react.

Starling realized that there were other instances of similar behavior, and coined the word "hormone" (from the Greek, meaning "to rouse to activity") to describe a substance released into the blood by one organ to prompt a response in another organ.

The work of Bayliss and Starling cleared the way for the recognition of diseases occurring from a hormone deficiency. Several years later another English physiologist, Edward Sharpey-Schafer, theorized that the pancreas produced a hormone that lowered the level of glucose in the blood. He named this hormone insulin, from the Latin word for island, as he believed it to be produced in the island cells of the pancreas. Within twenty years, the Canadian team of Frederick Banting and Charles Best devised a procedure for extracting a crude version of insulin from animals, and regular insulin treatments are now the standard method of controlling diabetes.

Banting devised the original experiments, and persuaded John Macleod, a physiology professor at the University of Toronto, to give him some laboratory space and find him a coworker. Macleod agreed, gave Banting the necessary space, and found Charles Best to work with him, and then promptly went off on a summer vacation. Banting and Best completed their work in 1922, and in 1923 Banting was one of two Canadians to share the Nobel Prize in medicine and physiology—the other being not Best, but Macleod! Banting was incensed, and almost refused to accept the Prize unless Best would share in it. He was unable to achieve this, but when he finally relented and accepted the Prize, he gave half of his share of the money to Best.

THE BIOCHEMISTRY OF METABOLISM

They met, fell in love, and married. Both were interested in the same aspect of science, and so they decided to work together. Because of the prejudice of the world of the early twentieth century against female scientists, she found it difficult to get employment in her chosen field. Nonetheless she persevered, and eventually managed to work alongside her husband. Their work was of such high quality that it was deemed worthy of a Nobel Prize.

It certainly sounds like the story of Pierre and Marie Curie, but it is also the story of Carl and Gerty Cori, whose lives (and names) parallel the Curies in many important aspects.

All living cells share certain characteristics. One of these is the ability to metabolize; to take in substances and use those substances to create both energy and new substances. One of the most important types of metabolism involves carbohydrates.

Over the past few decades, there has been a drastic change in the diet of an athlete prior to an important event. Athletes used to have steak and eggs; now they have pancakes or pasta. There is a sound reason for this. Pancakes and pasta are carbohydrates. Not only are they more quickly metabolized than protein-rich foods such as steak, but about half the carbohydrates are stored in the liver and muscles in the form of the chemical glycogen. The remainder is either stored as fat or burnt as fuel.

The biochemist Otto Meyerhof determined that, when a muscle contracts, the glycogen it has stored is converted to lactic acid (lactic acid is the source of the burning sensation to which the exercise instructor refers when he or she says, "Feel the burn!"). The lactic acid is then resynthesized into glucose, forming a cycle that enables the muscle to continue to contract. The work for which the Coris received the Nobel Prize took place over the course of many years, and consisted of a detailed analysis of this glycogen-to-lactic-acid-to-glycogen cycle.

While the Coris were working out what happened to the pancakes and pasta, the German biochemist Hans Krebs was doing the same thing for steak and eggs. Proteins are strings of amino acids, and Krebs discovered that metabolism removes the nitrogen atoms from the amino acids, eliminating them in the form of urea, the organic chemical first synthesized by Friedrich Wöhler.

The Coris, born in Czechoslovakia, had immigrated to the United States because of employment opportunities, but the rise of the Nazis in Germany compelled Krebs to immigrate to England. Like the Coris, Krebs became interested in carbohydrate metabolism. The Coris had shown that glycogen was converted to lactic acid without using oxygen, but released very little energy. Krebs decided that the remainder of the energy must be generated by chemical reactions that broke down the lactic acid and used oxygen, releasing water and carbon dioxide in the process.

This analysis took Krebs over five years to complete. One of the key intermediate products was citric acid, the same chemical that produces the slightly sour taste in orange or grapefruit juice. The cycle Krebs discovered is called the citric acid cycle, also known as the Krebs cycle. Later investigation revealed that the Krebs cycle is also responsible for the way in which fats are metabolized. The Krebs cycle is the major source of energy production in all living organisms.

The Curies were the first husband-and-wife team to win a Nobel Prize, and their daughter, Irene Joliot-Curie, was part of the second such husband-and-wife team. The Coris were the third, receiving their Nobel Prize in 1947. Since then, the environment for female scientists has improved substantially, and several have won Nobel Prizes—but no husband-and-wife teams. One reason might be that female scientists are much more numerous and much more a part of the scientific community than they were in the first half of the twentieth century, and so have a much wider choice of colleagues.

CHAPTER 9

Disease

Every so often, someone writes a book on science for the ages. *The Microbe Hunters*, written by Paul de Kruif and published almost one hundred years ago, is such a book. It describes the efforts of the early bacteriologists, including such legends as Louis Pasteur, as they landed the first real blows in the fight against disease.

Disease is the common enemy of every man, woman, and child—and it has only been in the last two or three centuries that we have come to recognize that disease is the result of natural causes rather than our having incurred the displeasure of the gods. Almost every generation for the last couple of centuries lives longer and enjoys better health than the generation that preceded it, and that is due in large measure to the efforts of the scientific and medical communities to understand, treat, and prevent disease.

Understanding Disease

It is impossible to imagine the terror that must have accompanied some of the great plagues of the past, such as the Black Death, which killed between 75 and 200 million people in Europe, Asia, and Africa between 1346 and 1353. In the fourteenth century, people had no idea what caused it, how to cure it, or how to prevent it. In 1894, Alexandre Yersin and Shibasaburo Kitasato independently discovered that the disease was caused by bacteria carried by fleas living on infected rats. There are few cases of bubonic plague today, but the antibiotic streptomycin, discovered in 1943 by a team of biochemists headed by Selman Waksman, is an effective treatment. Whenever a case of bubonic plague is detected today, the public is warned to stay away from a particular area, which generally harbors plague-infected rodents.

THE BIRTH OF EPIDEMIOLOGY: THE 1854 LONDON CHOLERA EPIDEMIC

John Snow is one of those individuals who deserves to be more widely recognized. However, if he were alive, he would probably not seek such recognition. Snow was a physician who was totally dedicated to his profession, spending a good portion of his career tending to the sick and injured of London's working class.

Not only was Snow a conscientious physician, he managed to stay apprised of the leading developments in medicine. Of course, doctors are supposed to do this routinely, but it is never easy. Communication in the nineteenth century was not what it is today, and even though the communication is instantaneous today, developments occur so rapidly that it is very difficult to keep up with them.

When Snow read of the use of the anesthetic ether in the United States, he realized that this was a development of profound significance. He studied the subject assiduously, and eventually wrote the definitive pamphlet, *On Ether*, concerning its use. When chloroform was introduced for childbirth a few years later, Snow was on top of that as well, and was the physician who administered the drug to Queen Victoria when she gave birth. As a result, Snow is recognized as the first anesthesiologist.

However, Snow's finest contribution to medicine came during the 1854 cholera epidemic that devastated London. It is difficult for us to imagine what the world must have been like before the introduction of antibiotics and chemotherapy. It was continually wracked by epidemics of diseases that are not a threat to westernized countries today, but that were capable of killing tens of thousands of people in the space of a few weeks. As COVID-19 has made abundantly clear, these still happen today, but not with the frequency that they did in the mid-nineteenth century.

In 1854 physicians had already noticed that the incidence of cholera was higher in dirty environments than in clean ones, but had leaped to the erroneous conclusion that the disease was caused by "miasmas," the foul odors associated with filth. Snow, on the other hand, believed that the disease was caused by the filth itself. Snow embarked on the first statistical analysis of an epidemic, keeping careful track of many items of data associated with sick and healthy individuals. He was able to show that areas supplied by water from the Southwark and Vauxhall Company, which got its water from the sewage-contaminated Thames, were nine times more likely to result in cholera fatalities than areas supplied by the Lambeth Company, which got its water upstream.

The most dramatic single bit of evidence concerned the Broad Street pump. Snow, who was familiar with the Soho area of London from his own practice, kept a complete record of the homes of those who perished of cholera. He noticed that more than five hundred fatalities occurred within a few hundred yards of the Broad Street pump. Snow discovered that a sewer pipe passed within a few feet of the well. After he managed to persuade the parish authorities to remove the pump handle, the fatalities were substantially reduced.

Although statistics had been invented more than two centuries earlier, Snow was the first to realize its potential value in medical applications. The man who was known as the first anesthesiologist was the founder of epidemiology as well.

Despite the fact that Pasteur had not yet established the germ theory, nor had Robert Koch shown that specific diseases were caused by specific organisms, Snow's experiences led him to believe that cholera was caused by a specific germ that lived and multiplied in water. With remarkable insight, he also recommended numerous public health procedures that are still standard practice: decontaminating soiled articles of clothing, washing hands, and boiling cooking utensils to sterilize them. Perhaps he was also the founder of public health.

THE GERM THEORY OF DISEASE

The great actors or actresses who spend a career in the theater usually follow a standard pattern. They start life with small parts and work their way up to supporting roles. Then comes a period of leading roles, followed by an inevitable regression into character parts. In a lifetime in the theater of science, Louis Pasteur never abandoned the leading role.

Pasteur began his career as a chemist. His first notable achievement was to demonstrate that different forms of crystals of the same compound were capable of polarizing light in different directions. His work on this subject sparked his interest in microorganisms, and the rest of his extraordinary career would be devoted to investigating these creatures.

In several instances, his investigations had profound economic effects. He discovered that the souring of both wine and milk was caused by microorganisms, and that this could be prevented by heating prior to bot-tling, a process now called pasteurization. His success was so remarkable that when the French silk industry was threatened by a disease afflicting silkworms, Pasteur was called on to pull another rabbit out of a hat. His investigations established that the disease was hereditary, and the silkworm colonies were then bred from eggs that had not been infected. This discovery saved the entire industry from ruin.

Of all his achievements, the one that had the most profound impact was his disproof of the theory of spontaneous generation. This theory held that microorganisms, some of which Pasteur had investigated while puzzling over the souring of wine and the disease of the silkworms, arose spontaneously in the substances in which they appeared. In an elegant series of experiments, Pasteur demonstrated that microorganisms were contained in the air, and that if substances were placed in a sterile environment and prevented from contact with air, microorganisms would not appear. This discovery gave considerable impetus to the germ theory of disease, and stimulated interest in microbiology.

Pasteur would not relinquish center stage. Fighting off the effects of a stroke that nearly killed him and left him partially paralyzed, he developed the theory of vaccine immunity, which held that it was possible to obtain protection from a disease by exposure to a weakened strain of that disease. In dramatic fashion, he demonstrated the validity of this theory by conducting an experiment in which he immunized twenty-five sheep against anthrax, and then injected them and twenty-five unvaccinated sheep with a strong dose of anthrax. All the vaccinated sheep lived, and all the unvaccinated ones died.

For his farewell appearance, Pasteur did what everyone said was impossible—he developed a vaccine against rabies that would enable indi-viduals who had been bitten by rabid animals to survive. For this he received international acclaim, and funds were solicited to build the Pasteur Institute in Paris for the investigation and treatment of disease. Many of the great names in bacteriology, and many of the important developments in the fight against disease, have come from the Institute.

Pasteur was not only a brilliant scientist and fervent patriot, but pos-sessed of traditional Gallic romanticism as well. After being made a profes-sor of chemistry at the University of Strasbourg, Pasteur fell in love with the daughter of the dean. He promptly wrote her a letter in which he said, "There is nothing in me to attract a young girl's fancy, but my recollections tell me that those who have known me very well have loved me very much. Time will show you that beneath an exterior cold and timid, which may displease you, there beats a heart full of affection for you." Searching for a comparison with which to convince the young lady of the depth of his feelings, he added, "I, who have been so in love with my crystals." This had the desired effect—Mademoiselle Laurent married him, and became not only his lifelong companion but doubled when needed as laboratory assistant, secretary, and coauthor.

THE DISCOVERY OF THE ANTHRAX BACILLUS

When Robert Koch met Emmy Fraatz, he was a romantic, a recently graduated doctor who wanted to roam the world, perhaps as a military physician or a ship's doctor. But Emmy was extremely practical, and since there would be little chance to raise a family onboard a ship, she persuaded Koch to enter private practice. They eventually settled in the small German town of Wollstein. Emmy recognized that Robert, who had graduated from medical school with highest distinction, felt the need of challenges beyond what a small-town medical practice would provide. Germany had become a leader in many industrial areas, including the production of precision scientific equipment. For Robert's twenty-eighth birthday, Emmy gave him a microscope.

Although nearly two centuries had passed since Anton von Leeuwenhoek discovered bacteria, medicine had made little progress. True, no reputable scientist believed that diseases were due to evil spirits, as was the case in Leeuwenhoek's day, but a fierce battle raged over two opposing points of view: the "miasmatic" theory, which held that disease arose spontaneously in individuals because of conditions in the external world, and the "parasitic" theory, in which disease was believed to be caused by microorganisms.

Perhaps because he was a country doctor, the disease that first commanded Koch's attention was anthrax, a disease contracted primarily by cattle and sheep. Livestock contracting anthrax died quickly and horribly; their blood turned a ghastly black. Koch examined this blood under a microscope, and observed that it contained stick-like organisms that often collected into long thread-like configurations. Others had conjectured that these organisms were the cause of anthrax, but it was Koch, in a brilliant series of experiments, who proved it.

Koch observed that these organisms were never found in healthy animals, but were always found in animals stricken with anthrax. He took blood from animals with anthrax and injected it into healthy mice. Within days, the mice contracted anthrax, and their blood contained the stick-like bacilli.

This proved that the blood of disease-ridden animals could be used to transmit the disease, but not that it was necessarily due to the bacilli, which might have been an effect rather than the cause. Koch therefore developed methods to grow pure strains of the bacilli, so he could be certain there were no other agents in the blood that might have been the cause of the disease.

More still remained to be done. Anthrax was obviously not transmitted from animal to animal by direct infection. Years of careful observation and experimentation were necessary to show that the bacilli changed into spores, which could live for long periods in fields. When the animals grazed on the spores, the spores would re-enter the bloodstream and change into the bacilli.

In the process of tracking down the cause of anthrax, Koch had not only shown that a specific microorganism was responsible for a specific disease, he established a methodology for the process of finding the causes of disease. This methodology became known as "Koch's postulates," and are still in use today.

It took some time for Koch's achievements to be recognized (he was, after all, an outsider as far as the medical establishment was concerned), but when they were, he became one of the first scientists to become a media celebrity. He then discovered, as other media celebrities would, that fame comes with a high price tag. His marriage ended in divorce, and when Koch married a much-younger actress, it was Koch's personal life that was placed under the microscope. Wanting only to continue his work, Koch became embroiled in the politics of medical research, and in later life fled to Africa, possibly as much to find peace and quiet as to investigate tropical diseases. With minor variations, the story of Robert Koch would be repeated a little more than half a century later when Jonas Salk developed a vaccine to prevent polio, and just recently, when Dr. Anthony Fauci became the face of the response to the COVID-19 pandemic.

Treating Disease

Curing disease has been one of the enduring concerns of mankind. All of us have fallen victim to disease at one time or another, and many apparently disparate cultures have discovered similarly effective ways of mitigating its effects. The benefits of chicken soup are extolled on every continent. While chicken soup may help assuage the miseries of the common cold, more serious diseases require more aggressive treatment.

THE DISCOVERY AND TREATMENT OF
DEFICIENCY DISEASES

By the middle of the eighteenth century, England's position as a world power had become inextricably intertwined with its ability to maintain a strong navy. In order to maintain a strong navy, it was necessary to find ways to keep a ship's crew healthy during long ocean voyages. A major problem faced by the British Navy was scurvy, a disease characterized by swollen and bleeding gums that weakened and then killed sailors.

James Lind began his medical career as a surgeon's mate in the British Navy, and immediately became interested in curing or preventing scurvy. After reading extensively on the subject, he became convinced that scurvy was the result of a monotonous diet lacking in fresh fruits and vegetables. He began treating sailors who had scurvy with an assortment of fresh fruits and vegetables, and discovered that the citrus fruits such as lemons and limes actually had the power to cure scurvy.

At the time, the British Navy was as bogged down by hidebound bureaucrats as many contemporary institutions. Lind persuaded Captain Cook to stock citrus fruits on his round-the-world voyage in the 1770s. Cook only lost one sailor to scurvy, but the British Navy remained unconvinced. Lind became the personal physician to King George III in 1783, but still could not persuade those in charge of the importance of citrus fruits in preventing scurvy. Lind died in 1794, but a year later the British Navy finally saw the wisdom of his idea, and adopted lime juice as a dietary staple, thus earning British sailors the nickname "limeys."

A century later, the brilliant discoveries of Louis Pasteur and Robert Koch had convinced scientists that diseases were caused by germs. Koch had been asked to go to the Dutch East Indies to find the cause of the disease beriberi. Koch was occupied investigating other diseases at the time, but recommended that Christiaan Eijkman, one of his students, work on the project.

The results were extremely disappointing, as the team was unable to isolate the germ responsible for beriberi, for the very good reason that no germ was in fact responsible. Most of the team returned, but Eijkman stayed on to become the head of a new bacteriological laboratory.

Some time later, a disease broke out among the laboratory chickens that looked suspiciously like beriberi. Once again, Eijkman tried to isolate the germ responsible, again with no luck. Then he tried another approach. He found that a cook had been feeding the chickens a diet of polished rice intended for hospital patients. The cook had been transferred, and the person who took over his job felt that chickens did not deserve to be fed specially treated rice, and so had resumed feeding them unpolished rice. Eijkman experimented along these lines, and discovered that diet was the difference— chickens fed on polished rice would develop the disease, but the disease could be cured by switching to unpolished rice.

The diseases for which Lind and Eijkman developed cures were deficiency diseases. Lind was the first doctor to develop a cure for a specific disease, and Eijkman was the first to discover that the absence of specific dietary ingredient was responsible for a disease. We are reminded daily of their contributions to our health, for the ingredient responsible for curing scurvy was vitamin C, and for curing beriberi, vitamin B.

A typical multivitamin tablet now includes almost half the alphabet. Initially these compounds were called "accessory food factors." However, when the chemist Casimir Funk investigated the compound responsible for curing beriberi, he discovered that it was an amine compound. He jumped to the conclusion that all of the accessory food factors *vital* to diet were *amine* compounds, and suggested the name "vitamine" to describe them. It was later discovered that not all such compounds were amines, and the terminal "e" was dropped.

ANESTHESIA

Of all the sciences, the one that is generally the most appreciated is medicine, because it most intimately touches our lives. There are very few developments in science that measurably affect the average human life span, but the discovery of anesthesia unquestionably belongs to this category. Without it, many of the operations that almost everyone will undergo at some point would be impossible.

Throughout history, man has sought relief from the tyranny of pain. Until quite recently, this relief could only be found either by the ingestion of large quantities of alcohol or through the consumption of substances such as morphine. While these achieve the effect of diminishing pain, they do it imperfectly and with many side effects. As a result, they are generally unsuitable for medical procedures, although many a tooth has been pulled or bullet removed after the consumption of significant amounts of alcohol.

The first great step toward the development of anesthesia was the discovery in 1800 of nitrous oxide by Sir Humphry Davy, the famous British chemist. Nitrous oxide exists as a gas, and its first "use" was at parties in nineteenth-century England known as "frolics." During these social occasions, Davy's gas was used to befuddle the participants. The gas was not only intoxicating, it made people hilarious—and soon came to be known as "laughing gas." It was noticed that people under the influence of "laughing gas" were temporarily insensitive to pain.

The colonists in the United States were also doing some experimentation along the same lines, although using different chemicals. The compound of choice for the Americans was ether, one of the oldest manufactured organic chemicals. Ether was undoubtedly developed by alchemists in the Middle Ages, who heated ethyl alcohol and sulfuric acid. In 1841, Charles Jackson of Plymouth, Massachusetts, discovered that ether had an anesthetic effect, although he made no immediate use of it.

Meanwhile, down in Georgia, the equivalent of the English "frolics" were taking place, using ether instead of nitrous oxide. During one of these parties, Crawford Long, a surgeon, realized the potential value of ether in medical procedures, and in 1842 removed a tumor from a patient's neck after first anesthetizing the patient. Although he used the procedure several times over the next few years, he did not publish his results until 1849.

In the meantime, Jackson had become acquainted with William Morton, a dentist who became interested in Jackson's observations concerning ether. In 1846, in consultation with Jackson, Morton administered ether to a dental patient and then removed a tooth. He also removed a tumor from another patient's neck, and published the results. Jackson and Morton applied for a patent on ether, but spent much of the rest of their lives quarrelling over the credit for discovering anesthesia. Crawford Long later felt that he should be similarly recognized, and this undoubtedly motivated his 1849 publication of his 1842 operations.

Alcohol, nitrous oxide, and ether may have been the first "recreational" drugs, but the relationship between such drugs and useful medical compounds continues to this day. Opium led to the development of such useful opiates as morphine, and the novocaine that is often administered in a dentist's office is very closely related to cocaine. Even cannabis is used in some instances as a treatment for glaucoma.

SALVARSAN AND SYPHILIS

One of the most exciting feelings a scientist can experience occurs when he or she gets an idea that no one has had before. Early in his career, Paul Ehrlich, who was an inveterate reader, had seen an article about lead poisoning in dogs, in which the author had determined that different amounts of lead accumulated in the tissues of different organs. This explained why lead had a more toxic effect on some organs than others.

This led Ehrlich to develop the revolutionary idea of *chemotherapy*: that a disease could be treated by finding a substance that was toxic to specific disease-causing organisms. In order to begin work, Ehrlich made a hypothetical connection between ideas he had come across in two other articles. The first article stated that trypanosomes, which were responsible for diseases such as African sleeping sickness, and spirochetes, which were responsible for syphilis, were closely related. The second article showed that atoxyl, an organic arsenic compound, had a toxic effect on trypanosomes.

As a result, Ehrlich formulated the hypothesis that an arsenic compound could be developed to treat syphilis. This idea, so sensible in retrospect, was regarded within the medical research community as laughable at best and dangerous at worst. Ehrlich, however, was undeterred. By 1905, when he began his search for a "magic bullet" to kill the spirochetes that caused syphilis, the synthetic chemical industry had advanced to a point where it was possible to generate numerous variations on the atoxyl theme.

One can only wonder what would have happened if Ehrlich had been employed, not by a German university at the turn of the twentieth century, but by a modern pharmaceutical company with its emphasis on the bottom line. For after one year of product development, no product had been developed. Nor after two years, nor three. More than four years and over six hundred unsuccessful trials took place before Ehrlich's Compound Six-Oh-Six (later renamed Salvarsan), the 606th atoxyl variant to be tested, proved successful.

Salvarsan proved to be successful beyond even Ehrlich's expectations. In many cases, it cured syphilis overnight. Ehrlich, whose modesty was legendary, was once congratulated on his achievement by a colleague. He replied that "for seven years of misfortune I had one moment of good luck."

Ehrlich was perhaps even luckier with Six-Oh-Six than he knew. It was later discovered that the cultures of syphilis continue to thrive almost normally on culture plates in laboratory experiments when exposed to concentrations of Salvarsan much greater than could possibly be given to patients. From other evidence, we now also know that the complex arsenic compound Ehrlich invented does not actually have the "magic bullet" effect for which he searched. It breaks down in the body to a simpler substance that actually produces the cure. If these experiments on culture plates had been made by Ehrlich, the first magic bullet to combat bacterial infections might never have been found. Perhaps fortune really does favor the bold, even in science.

Many of Paul Ehrlich's personal characteristics typified what we have come to think of as the absentminded professor. He thought nothing of scribbling chemical formulas on any available writing surface, including those not expressly designed for the purpose, such as linen tablecloths. Despite numerous personal idiosyncrasies, almost everyone found him a man of great warmth and personal charm. When he died in 1915 during the First World War, even as British and German troops were facing each other throughout Europe, the *London Times* eulogized him: "He opened new doors to the unknown, and the whole world at this hour is his debtor."

THE DISCOVERY OF PENICILLIN

Discoveries in science are supposed to go "through channels"; they are written up, submitted to professional journals, undergo peer review, and are then published, assuming they meet the standards required for publication. A scientist can also publicize his or her discovery within the confines of the scientific community, usually by giving lectures or seminars. Indeed, if a scientist feels that the discovery is noteworthy, it is important that he or she make personal appearances to bring the discovery to the immediate attention of the scientific community. But it doesn't hurt to be a good promoter— which Alexander Fleming was not.

In 1921, Fleming found himself a victim of what was probably a common cold. One day, while working with his bacterial cultures, some of the mucus produced by this cold accidentally dripped into one of the culture dishes. Examining this culture a few days later, Fleming noticed that the bacteria had been destroyed; the liquid had been turned, as he later said, "clear as gin." He isolated the substance responsible for this phenomenon and named it lysozyme. It turned out to be an enzyme that exists in tears, saliva, and mucus.

Feeling that this was an important discovery, he arranged to deliver a lecture before the prestigious London Medical Research Club. Unfortunately, Fleming was a quiet, soft-spoken man who usually said little. His strength was that of a keen observer, rather than a speaker or expositor. As a result, when he concluded his lecture, his audience reacted with total indifference.

Seven years were to pass. In the summer of 1928, in the process of some rather routine research on the growth and properties of staphylococcal bacteria, Fleming prepared a number of culture dishes, and covered them with glass plates. He then went on vacation. Upon returning, he discovered that one of the glass coverings had fallen off a dish. One of the molds that are always present in dusty laboratories had gotten on to the bacteria dish and "spoiled" it. He was just about to throw out the contaminated dish, when he suddenly changed his mind. He examined the dish more closely and noticed that a gaping hole appeared in the middle of the culture. Fleming analyzed it and discovered that it was due to an accidental contamination of the dish by a fungus. This mold was later identified as *penicillium notatum*, whose active ingredient was penicillin. Fleming, calm and even-tempered as always, duly recorded this observation. Despite the chilly reception he had endured from the London Medical Research Club eight years previously, he felt his discovery was sufficiently important to report it to them.

Once again, the enormous promise of Fleming's discovery was obscured by his inability to convince his audience of the importance of his discovery. It required the stimulus of World War II, with its urgent need for antibiotics, to recognize Fleming's discovery for what it was: the single most powerful weapon ever developed in the battle against microbial infection, a drug that would save tens of millions of lives.

The value of penicillin was recognized by the beginning of World War II, but there were major obstacles to mass-producing it. The British pharmaceutical industry did not have the time or the resources to pursue the problem, so the world's entire supply of penicillin, which consisted of a few test tubes derived from Fleming's original discovery, was packed up and shipped in secret to the United States. It was believed to be perhaps the most valuable substance on the planet, and every single milligram was jealously hoarded. When a laboratory technician at Merck, one of the drug companies investigating ways to mass-produce it, requested a larger sample of penicillin for an experiment, Dr. Max Tishler, the laboratory director, responded, "Remember, when you are working with those 50 or 100 milligrams, you are working with a human life."

Preventing Disease

As of this writing, there is no cure for COVID-19—although there are developments that show promise. But there is something even better than being able to cure a disease, and that is being able to prevent a disease—or to minimize its effects should it occur.

If the germ for a disease exists, there are people whose immune system will protect them from contracting the disease, but the immune systems of many people do not recognize the germ as hostile and thus do not mobilize the body's defenses to combat that disease. But one of the great discoveries in the history of science is that there is a way to get the body's immune system to recognize the germ without incurring the cost of first contracting the disease—and that's what vaccines do.

THE SMALLPOX VACCINE

A fierce debate has raged over the years on whether or not to execute one of the greatest mass murderers of all time. This mass murderer remains under constant guard in the Centers for Disease Control and Prevention in Atlanta, Georgia, as well as in a similar laboratory in Russia. The debate centers around the last remaining specimens on Earth of the smallpox virus, and the question is whether it would do more good to have it available for study, or whether it should be destroyed to prevent it from ever again afflicting mankind.

The last recorded case of smallpox was in 1977, and nowadays everyone is routinely vaccinated against it. Since almost no one has even seen a case of smallpox, we must rely on descriptions such as the following by the noted author Thomas Macaulay:

> That disease . . . was the most terrible of all the ministers of death
> . . . the smallpox was always present, filling the churchyard with
> corpses, tormenting with constant fears all whom it had not yet
> stricken, leaving on those whose lives it spared the hideous traces
> of its power.

Toward the end of the seventeenth century, the Turks had noticed that those who survived an attack of smallpox developed immunity to the disease. They developed a technique of smallpox inoculation, in which a person was given a (hopefully) mild case of smallpox. Unfortunately, this technique, called variolation, was highly erratic. If the individual being inoculated contracted too severe a case, he or she could easily be scarred for life, blinded, or even killed.

Dr. Edward Jenner, an English country physician, was familiar with the technique of smallpox inoculation. While inoculating villagers one day, he was told not to bother inoculating one of the dairymaids. When Jenner asked why, the villagers explained to him that she had previously contracted a case of cowpox. While it was known that people who had contracted cowpox never got smallpox, it was Jenner who formulated the critical hypothesis: if one were to give people the mild disease of *cowpox* via inoculation, rather than the dangerous disease of *smallpox*, one could achieve the immunity against smallpox without risk.

Jenner took more than twenty years of experimentation to establish the truth of this hypothesis. In the process, he coined the term "vaccine" (from *vacca*, the Latin word for cow!). Nonetheless, when he presented his findings to the Royal Society, he admitted that his investigations were incomplete, stating that he hoped his results would "present to persons well situated for such discussions, objects for a minute investigation. In the meantime, I shall myself continue to prosecute this inquiry, encouraged by the hope of its becoming essentially beneficial to mankind."

It would be more than a century before the mechanism by which smallpox vaccination worked would be understood. The immune system stores a record of previous invasions by foreign bodies, so that future onslaughts may be quickly repelled. Once the immune system has encountered the cowpox virus, it is able to recognize and attack the highly similar smallpox virus before the latter has time to multiply and overwhelm the body's defenses. Encouraging the immune system to destroy an invading virus is the basis of all vaccines, including the vaccines recently developed against COVID-19.

Many of the great advances in science are the result of acute observation. Good observers are doubtless made, not born, but there are probably certain backgrounds that prepare one well for patient observation. Jenner found an unusual one, as prior to becoming a doctor, he had been an ornithologist. His chief claim to fame had been the observation that cuckoo chicks, which hatch from eggs laid in other birds' nests, physically eject the natural chicks from their own nest, and are then "adopted" by the birds who raise the cuckoo chick as their own.

ANTISEPSIS

Modern surgery, as it is currently performed in hospitals, owes its high rate of success to two dramatic developments that occurred during the nineteenth century. The first was the introduction of anesthesia, which made it possible to perform surgical procedures of extended duration. The second was the institution of antiseptic precautions on the part of the surgeons, which greatly reduced post-operative mortality.

Ignaz Semmelweis was a Viennese pre-law student who accompanied a friend to an anatomy lecture. Anatomy impressed Semmelweis as being far more interesting than law, and he eventually received a medical degree from the University of Vienna. On graduation, he immediately became interested in childbed fever. He was startled by the fact that women who were examined by a physician prior to childbirth were much more likely to die of infection than those who were not examined. When a colleague died of an infection after being cut by a surgical knife, Semmelweis reached the conclusion that lethal infections were being caused by unsterile instruments and physicians' hands. Against fierce opposition, he required doctors in his department to wash their hands prior to performing surgery. This resulted in a dramatic reduction in the number of women who died from post-childbirth infection. Despite these clear lifesaving benefits, Semmelweis was fired from the hospital, and was persecuted to such an extent that he suffered a nervous breakdown. In a sadly ironic twist, Semmelweis suffered the same fate as his colleague. While attending a sick patient, he accidentally wounded himself and died of childbed fever.

Several years later a Scottish physician, Joseph Lister, became convinced of the validity of Louis Pasteur's theory that microbes caused infections. The introduction of anesthetic procedures had enabled doctors to perform more complex surgeries, but the gains made thereby were offset by a stunning increase in gangrene and similar infections. Building on Pasteur's ideas, he decided to institute measures designed to kill any germs that might exist in surgical wounds by treating them with carbolic acid. Although this had the unpleasant effect of irritating the treated tissues, it greatly reduced post-surgical infections. The reaction to Lister's efforts by his surgical colleagues paralleled the experiences of Semmelweis. The surgeons in Lister's hospital rejected his methods and refused to adopt his procedures. He was accused of stealing the ideas of others, and had to struggle for more than a decade before his antiseptic procedures finally gained credence. Although he did his initial work in Scotland, he later moved to London in an attempt to convince surgeons there of the validity of his techniques. His ideas met with stiff opposition in England, but they were quickly adopted on the continent.

Lister was not only an excellent surgeon, he was a respected scientist who stayed on the cutting edge (appropriate for a surgeon) of developments in his field. He was one of the first scientists to accept Pasteur's and Koch's theories, and did pioneering work in bacteriology.

Even though Semmelweis and Lister were basically attacking the same problem, Lister had the enormous advantage of timing. The sterilization techniques introduced by Semmelweis were successful, but at the time there was no apparent reason why they should have been. Lister's work occurred at the time that both Pasteur and Koch were achieving great recognition, and in the light of these developments it was easy to understand why Lister's techniques proved so effective. It would be nice to think that the scientific world is one in which rationality always prevails, but the experiences of Lister demonstrate that just because what you do works, and there is a valid reason that it works, there is no guarantee that your ideas will be immediately accepted.

The admonition to "wash your hands thoroughly" came once again to the fore during the COVID-19 pandemic. This antiseptic procedure helps prevent both bacterial and viral infections by removing or killing the germs responsible. Germs generally enter the body through openings, such as eye, nose, and mouth, that we touch frequently. Even though it has been shown that COVID-19 is primarily spread through respiratory droplets, it's still a good idea to substitute the fist or elbow bump for the common handshake during the pandemic.

THE POLIO VACCINE

As of this writing, the battle against COVID has been going on for several years, and progress has been remarkable. Such was not the case in 1949, when the dreaded disease was not COVID-19 but polio. Like COVID, polio was caused by a virus. Unlike COVID, whose victims tend to be the elderly, polio primarily attacked children, and was also known as infantile paralysis. It was not known precisely which behavior was most likely to result in a person contracting polio, but the disease altered behavior much as COVID has. During the summer, parents forbade their children from swimming in public pools, and people slept with their windows closed to prevent the virus from entering. Leading experts agreed that a cure or a vaccine was decades away.

The experts were wrong. On April 12, 1955, ten years to the day after the death of President Franklin D. Roosevelt, a well-known polio victim, it was announced that a vaccine developed by a young medical investigator named Jonas Salk, from the University of Pittsburgh, had been proven safe and effective in preventing polio. The efficacy of the Salk vaccine had been confirmed in one of the largest field trials in medical history. Had Salk discovered that chicken soup prevented polio, the news would have been no less welcome, and the road to eventual success would have been substantially easier, for no expert's reputation depended upon the denial of the curative power of chicken soup. Salk, however, decided to try to create a "killed virus" vaccine. At the time, the prevailing view of vaccine immunity was that it was necessary to use "live virus" vaccines, of the type developed by the nineteenth-century giants Louis Pasteur and Robert Koch. In trying to create a "killed virus" vaccine, Salk was challenging the wisdom of the medical establishment. Salk's success was to be purchased at the high price of hostility from many of his peers in medical research.

Salk had first come to doubt the validity of the "live virus" theory in 1936, while a twenty-one-year-old medical student at New York University. During a lecture on immunization, he had heard the professor reiterate Pasteur's theory that immunization against viral diseases required live viruses. Yet, in another lecture, he also learned that the vaccine against diphtheria, a bacterial disease, was created from *killed* diphtheria toxin. Salk was puzzled by the paradox. Why should a vaccine against viral diseases be so radically different from those successfully employed in the fight against bacterial diseases?

With the advantage of hindsight, it is somewhat difficult to see why a killed virus vaccine was such a heretical idea. Isabel Morgan, niece of the famous geneticist Thomas Hunt Morgan, had shown that a killed virus vaccine could be developed to immunize monkeys against polio. At the time it was thought that monkeys were so radically different from man that this research had no bearing on the development of a vaccine for human beings. DNA analysis performed during the 1980s has since shown that the chemical structure of the blood of man and chimpanzee matches to within 1 percent. Had this been known at the time, perhaps Salk's strategy would not have appeared so controversial.

Perhaps never in the history of science has an individual who did so much good been so excoriated by his peers. The adulation heaped upon Salk, which Salk tried desperately to avoid, created an unparalleled enmity among his colleagues. Salk saw his achievement as the culmination of decades of work by many scientists, and repeatedly tried to present that view, but the public and the media didn't buy it. Some of his colleagues saw Salk as committing the three cardinal sins of a scientist: challenging the wisdom of the establishment, being correct in doing so, and becoming famous as a result. Despite his monumental achievement, he never became a member of the prestigious National Academy of Sciences—but there are a number of Salk Institutes and schools named in his honor, and almost none for members of that academy.

CHAPTER 10

Science in the Twenty-First Century

We are only two decades into the twenty-first century, and already we have seen some amazing discoveries. Gravitational waves, predicted by Einstein's theory of relativity, have not only been discovered but have shed light on what happens when collisions occur between massive objects such as neutron stars. The discovery of the Higgs boson, which gives mass to the various particles, was finally confirmed after a half century of searching. And, of course, the rapidity with which the mRNA vaccines were created and produced have helped to mitigate the COVID-19 pandemic, and hold the promise of revolutionizing vaccine creation.

As amazing as these discoveries are, the promise of even more lies ahead. The Webb telescope may show us the signature of life elsewhere in the Universe. We may finally discover the nature of the dark matter and dark energy that appear to comprise most of the matter and energy in the Universe. We may discover how life evolved on Earth, and there is a good chance that this century will see the creation of life-forms from scratch, rather than by modifying existing life-forms.

Already we can see three developing trends in science. What has changed in the twenty-first century is how science is done, and who's doing it. Sadly, what has not changed are the negative and erroneous views with which science is sometimes regarded.

HOW SCIENCE IS DONE IN THE TWENTY-FIRST CENTURY

When we look at the milestones of science, the vast majority to date are the result of the work of individuals. For most of the seventeenth and eighteenth centuries, scientific developments occurred as the result of the investigation or theorizing by a single individual. Granted, each individual did have the ability to correspond with other scientists through the mail, and they might even occasionally meet, but most of the work was done in isolation. Books or articles would be published and read, but there would often be delays of weeks or months before other scientists learned of new developments. Almost all scientific discoveries took place in Europe, as communication between Europe and the Americas required ocean voyages. For the most part, science was pursued by men who had the financial resources and time to pursue it.

The nineteenth century saw the rise of scientific collaboration. Nations and industries saw the advantages that scientific developments brought, and investment in science increased. Publishing became more widespread, and railroad trains made travel substantially easier. As a result, scientists could meet with others and learn of their work fairly rapidly. The speed of discovery increased.

This trend accelerated during the twentieth century. Electronic communication made it possible for scientists to learn of new developments almost instantaneously. The airplane made it possible for large groups of scientists to easily congregate, and scientific conventions attended by hundreds or thousands of scientists became usual events. Science became a major beneficiary of government funding. Additionally, public awareness of science increased immeasurably, and more people sought scientific careers in industry, government, and academia.

The importance of the electronic computer to scientific progress cannot be underestimated. In the middle of the twentieth century, a department in a university or a business might have a single computer. By the end of the twentieth century, it was common for every faculty member to have his or her own computer. Although large-scale databases were not available to the extent they are now, email enabled scientists to communicate with each other overnight. The computer was capable not only of solving difficult problems, but simulating situations that could not possibly be studied any other way.

These trends have, not surprisingly, continued. Major developments are known instantaneously, and are often the result of large-scale collaborations between scientists in many nations. This has been made easier by the de facto adoption of English as the universal language of science. Additionally, platforms such as Zoom have made it possible for scientists in one portion of the world to visualize an experiment as it is taking place half a world away. In the nineteenth century, scientists on different continents could write to each other. In the twentieth century, they could talk to each other via telephone. Now their computers enable them to see and actually work with each other in real time. It's the Golden Age of scientific communication and collaboration.

WHO'S DOING SCIENCE IN THE TWENTY-FIRST CENTURY

The short answer is—practically anyone that wants to.

Again, perusing the history of science makes it apparent that during the seventeenth century, most of science was done by men—and European men in particular. In the nineteenth century, science became more of an international activity, and women started to contribute to the advancement of science. Science was still regarded as an activity mostly pursued by men.

That began to change in the twentieth century. More women entered scientific fields. Even though some women were recognized—Marie Curie received two Nobel Prizes—some were ignored or suppressed, as we recall from the story of the Nobel Prize Committee not realizing that Henrietta Swan Leavitt had died some five years prior to their consideration of her. But, as gender and civil rights movements made social progress, their impact was felt in the sciences.

Nowadays, women are a significant part of the scientific workforce, and are likely to become even more so in the future. The National Science Foundation compiled statistics, and currently 43 percent of the scientific workforce is female, and that percentage rises to 56 percent if one considers only scientists under twenty-nine years of age.

Science is much more of an international effort than ever before. Most governments support scientific endeavors, and—as mentioned previously—the internet enables instantaneous communication between any two scientists anywhere in the world. I've collaborated with scientists from Europe and Asia whom I've never met—and probably will never meet—and this is the rule rather than the exception.

Additionally, it is now possible for the public at large to make contributions to scientific research. The widespread availability of scientific data has made it possible for amateurs to make useful contributions. In 2021, two high school students combing through data discovered four exoplanets. It would not be surprising if similar discoveries were made by amateurs sifting through the mass of data currently made available through various genome projects.

Developments such as these are extremely heartening. Scientific knowledge should be available to everyone, and everyone should be able to contribute to amassing it and gleaning useful results therefrom. It is not only beneficial; it is intellectually satisfying, and helps bring people closer together. Everyone wins.

HOW SCIENCE IS PERCEIVED

Carl Sagan, an internationally respected scientist, was one of the great popularizers of science in the twentieth century. His book *Cosmos* was made into an extremely popular television miniseries, and Sagan himself was a staple on television talk shows. Another of his books, *The Demon-Haunted World: Science as a Candle in the Dark*, is a stark description of the dangers presented by the antipathy to scientific thought that has existed throughout history.

Sagan would not have been at all surprised to learn that fully 20 percent of Americans believe that the government is injecting tracking microchips via the COVID vaccines. Here we are, in the midst of the first global pandemic in a century. The population has been completely vaccinated against smallpox and polio—and also receives periodic vaccination against tuberculosis, shingles, and flu. Nonetheless, certain parties have found it of greater value to spread lies about the mRNA vaccines and the motives behind those urging vaccination than to implore their followers to take advantage of one of the great achievements of medical science.

Science has faced many uphill battles in the past. In the Middle Ages, scientists such as Giordano Bruno were burned at the stake for promulgating the heretical notion that the Earth was not the center of the Universe. Such a point of view brought science into confrontation with the Roman Catholic Church, the most powerful organization of the times. Most of the battles science now faces are fought on the field of the biological sciences. It's hard to imagine a development in physics or chemistry that would engender societal conflict—and it is hard to believe that the development of the mRNA vaccines to ease a global pandemic would have done so. But it did, and unless something is done in the way of counteracting the inaccurate information that proliferates via social media, things will probably get worse.

Science is one of the great benefactors of humanity, but there is no question that science opens Pandora's box. Almost certainly this century will see the creation of life from test tube ingredients—and given the array of DNA analysis and engineering tools currently in existence, we will see the opening of a Pandora's box unlike anything we have ever witnessed. Let us hope that we have the wisdom to accompany the knowledge that science has accumulated and will accumulate.

Bibliography

Asimov, Isaac. *The Universe*. New York: Discus Books, 1971.

Broda, Engelbert. *Ludwig Boltzmann: Man, Physicist, Philosopher*. Woodbridge, CT: Ox Bow Press, 1983.

Bronowski, Jacob. *The Ascent of Man*. New York: Little, Brown, 1973.

Burke, James. *Connections*. New York: Simon and Schuster, 2007.

Carson, Rachel. *Silent Spring*. Boston: Houghton Mifflin, 1962.

Christianson, Gale. *Edwin Hubble: Mariner of the Nebulae*. New York: Farrar, Strauss, Giroux, 1995.

Crosland, Maurice P. *Gay-Lussac: Scientist and Bourgeois*. Cambridge: Cambridge University Press, 1978.

Dawkins, Richard. *The Blind Watchmaker*. New York: W. W. Norton, 1986.

de Kruif, Paul. *The Microbe Hunters*. New York: Harcourt, Brace, 1926.

Einstein, Albert. *The Evolution of Physics*. New York: Simon and Schuster, 1961.

Gamow, George. *One, Two, Three . . . Infinity: Facts and Speculations of Science*. New York: Dover, 1947.

Gillmor, Stewart. *Coulomb and the Evolution of Physics and Engineering in Eighteenth-Century France*. Princeton, NJ: Princeton University Press, 1971.

Greene, Brian. *The Elegant Universe: Superstrings, Hidden Dimensions, and the Quest for the Ultimate Theory*. New York: W. W. Norton, 1999.

Gribbin, John. *The Scientists: A History of Science Told through the Lives of Its Greatest Inventors*. New York: Random House, 2003.

Hart, Michael. *The 100: A Ranking of the Most Influential People in History*. Secaucus, NJ: Carol Publishing Group, 1987.

Hellemans, Alexander. *The Timetables of Science*. New York: Simon and Schuster, 1988.

Holmyard, Eric J. *Makers of Chemistry*. Oxford: Oxford University Press, 1953.

Livingston, Dorothy. *The Master of Light*. New York: Charles Scribner and Sons, 1973.

Oberg, Erik. *Machinery's Handbook*. 29th edition. New York: Industrial Press, 2012.

Sagan, Carl. *Cosmos*. New York: Random House, 1980.

Scientists Who Changed History. London: D.K. Publishing, 2019.

Staley, Richard. *Einstein's Generation: The Origins of the Relativity Revolution*. Chicago: University of Chicago Press, 2008.

Sykes, Christopher (Ed.). *No Ordinary Genius: The Illustrated Richard Feynman*. New York: Norton, 1995.

Wali, Kameshwar. *Chandra: A Biography of S. Chandrasekhar*. Chicago: University of Chicago Press, 1990.

Watson, James. *The Double Helix*. New York: Scribner, 1998.

Weast, Robert (Ed.). *CRC Handbook of Chemistry and Physics (1981–1982)*. Boca Raton, FL: CRC Press, 1981.

Acknowledgments

There are several people that I'd like to thank who helped make this book possible. First up are my parents, whose obvious contribution was supplemented by encouraging my love for science. My parents took me to museums, gave me presents such as a cherished chemistry set, and checked out books that they felt might be of interest from the local library. Next is my wife Linda, who returned from a trip to Taiwan just in time to realize that what I thought was a cold might be more serious. It was pneumonia, and without her efforts I might not be here to write this book. The third person I would like to thank is my editor, Jake Bonar, for realizing that this might be just the right time for a book like this, and I might be the right person to write it. Last, I am grateful to Al Posamentier, a fellow mathematics professor whose connections in the publishing world fortunately included Jake, and who brought the two of us together.

Index